Nick Vandome

Video Chatting
for Seniors

004.695 Van

In easy steps is an imprint of In Easy Steps Limited
16 Hamilton Terrace · Holly Walk · Leamington Spa
Warwickshire · United Kingdom · CV32 4LY
www.ineasysteps.com

Notice of Liability
Every effort has been made to ensure that this book contains accurate
and current information. However, In Easy Steps Limited and the
author shall not be liable for any loss or damage suffered by readers
as a result of any information contained herein.

Trademarks
Microsoft® is a registered trademark of Microsoft Corporation.
iPhone®, iPad®, OS X® and macOS® are registered trademarks of Apple
Computer, Inc.
All other trademarks are acknowledged as belonging to their
respective companies.

In Easy Steps Limited supports The Forest Stewardship Council (FSC),
the leading international forest certification organization. All our titles
that are printed on Greenpeace approved FSC certified paper carry the
FSC logo.

MIX
Paper from
responsible sources
FSC
www.fsc.org   FSC® C020837

Printed and bound in the United Kingdom

ISBN 978-1-84078-932-4

Contents

1 The Video-Chatting Revolution

This chapter introduces the concept of video chatting and some of the issues involved.

Video Chatting Comes of Age

Change happens all around us all of the time, particularly in the digital world, where technology seems to change and develop at a breakneck speed. Sometimes changes in the digital world can appear to be suddenly with us as if they were always there, while at other times there is a watershed moment that embeds the technology into our daily lives. For video chatting, that watershed moment was the Covid-19 pandemic, which has dramatically transformed the way that people communicate with each other. The change has been so significant that it will undoubtedly ensure that video chatting is now a constant part of millions of people's lives.

Although video chatting has been around for a number of years, it was the global lockdowns as a result of Covid-19 that made it an indispensable tool for family and friends to keep in touch. Being able to talk to people – and see them – has been a vital link for millions of people and has increased the profile of video-chatting apps significantly.

Benefits of video chatting

It is not hard to see why video chatting has become so popular, due to the number of benefits it brings:

- **Cost**. Video chatting can be done on devices that a lot of people already have: tablets, smartphones, laptops and desktop computers. The apps for conducting video chats all have free versions, so you can get up and running without any additional costs.

- **Versatility**. Most video-chatting apps are multi-platform, which means they can be used on a range of devices, using different operating systems.

- **Mobility**. Since video chatting can be used on smartphones, this means that you can conduct a video chat wherever you are, with a smartphone. It is common to see people using smartphones to have video chats rather than for making voice calls.

- **Effectiveness**. Video chatting apps are excellent at what they do, making the experience both effective and fun.

Don't forget

The main video-chatting apps looked at in this book are Zoom, Skype, FaceTime and Messenger.

Don't forget

Making video calls to other people using the same video-chatting app is free over Wi-Fi, and it can also be done with a data connection on a smartphone, although data charges may apply for this, depending on your smartphone contract provider.

About Video Chatting

Video chatting can be thought of in the same way as making a voice call: an appropriate app is used to connect two – or more – people who want to take part in the chat, and they can do this on their favored computing device. A Wi-Fi connection is also needed for video chatting (or a mobile data connection for a smartphone), since the connection between the participants is done over the internet. So, to get started with video chatting:

See Chapter 2 for more details about the equipment, connections and apps that can be used for video chatting.

1. Use your favored computing device and download a video-chatting app, such as Zoom, Skype or Messenger

2. Connect your device to your home Wi-Fi

3. Open a video-chatting app

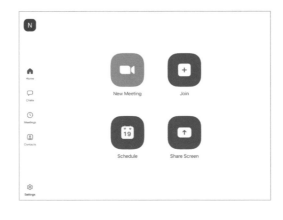

Apple devices, such as the iPhone and the iPad, contain the FaceTime app for video chatting. However, this can only be used with other FaceTime users on an Apple device. Multi-platform apps such as Zoom, Skype and Messenger can also be used on Apple devices for video chatting.

...cont'd

4 Use the **Contacts** section on your video-chatting app to view people who are available for video chats, and invite more as required

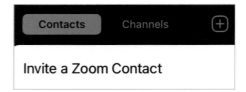

In most video-chatting apps, contacts can be imported from the address book (usually the **Contacts** app) on the device on which the video-chatting app is being used. Video chats may also be arranged via text messages.

5 Add more people to the **Contacts** section of your video-chatting app so that you can select them from here when you want to chat

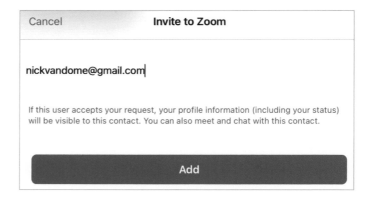

Video chats are often referred to as meetings in video-chatting apps.

6 Start a video chat (or meeting) with one of your contacts in the video-chatting app

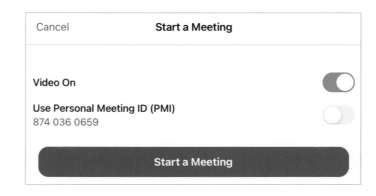

7 Once a one-to-one video chat has been connected, your own video feed is displayed as a thumbnail in a corner of the screen, and the other person's video feed is displayed in the main window

8 Add more people to the chat to create a group chat

Beware

The greater the number of people in a group call, the harder it can be to control it effectively.

...cont'd

9 Customize your video feed with creative backgrounds (depending on the video-chatting app being used)

Beware

Not all video-chatting apps and devices have the functionality for adding special effects.

10 Customize your video feed with special effects (depending on the video-chatting app being used)

11 Schedule recurring video chats at a specific time – e.g. once every week for a family chat or a group quiz

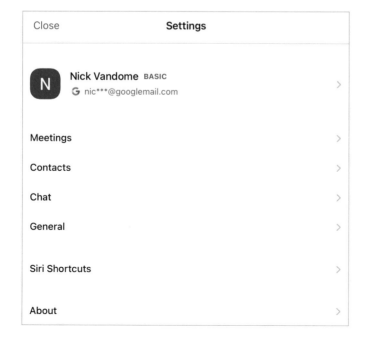

12 Use the Settings options for the video-chatting app to customize it to your own requirements

Hot tip

Investigating the Settings options of a video-chatting app is a good way to discover its features and functionality and to also become more confident when using the app.

Cross-Platform Performance

Since there are multiple types of devices and different types of operating systems that can use video-chatting apps, it is important that they can be used on as many devices as possible so that you can switch between devices if required, and still use the same app. Using apps in this way is known as operating cross-platform.

The types of devices that can use video-chatting apps, and their respective operating systems, are:

- Desktop and laptop computers using the Windows 10 operating system.

- Desktop and laptop computers using Apple's macOS operating system.

- Smartphones and tablets (iPhone and iPad) using Apple's iOS and iPadOS operating system respectively.

- Smartphones and tablets using the Android operating system.

In terms of their cross-platform performance, the main apps in this book can be used as follows:

- **Zoom**. This is highly versatile and can be used with all major operating systems and devices. The interface is generally similar between all versions.

It is necessary to create a user account to use a video-chatting app. Once this has been done, you can download the app onto any device and use it with your account login details (apart from FaceTime, which can only be used on Apple devices).

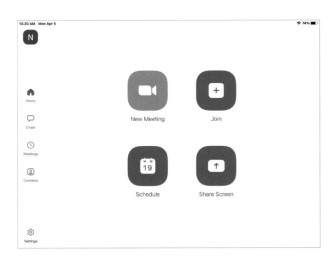

- **Skype**. This can also be used with all major operating systems and devices, although the desktop and laptop versions (below, left) have a slightly different interface from the mobile versions (smartphone and tablet – below, right). All versions have the same functionality, though.

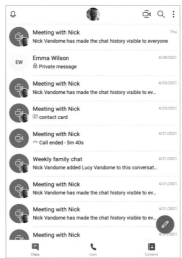

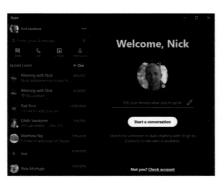

Don't forget

One obvious difference between desktop/ laptop versions of video-chatting apps and their mobile versions is the available screen size. Because of this, the mobile versions are more compact, with toolbars at the top and bottom of the screen to access the same functionality as the desktop/laptop versions of the apps.

- **FaceTime**. This is Apple's own video-chatting app and it can only be used on Apple's own devices. The main difference between devices is that the iPhone version has a button for adding special effects.

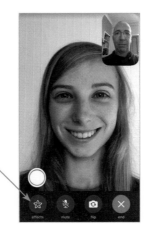

- **Messenger**. This is another video-chatting app that is highly versatile and works equally well with different operating systems and devices.

In general, you do not have to worry about cross-platform issues with video-chatting apps.

More Than One-to-One Video

Video-chatting apps are excellent at performing one of their main tasks – i.e. providing video communication between two people. However, most video-chatting apps are much more versatile than this and offer a much wider range of communication options. These include: making audio calls (that can also quickly be converted into video calls); making group calls; text chatting; and sharing content in a call.

Audio calls

Video-chatting apps are just as proficient at making audio calls as they are at making video ones. The main difference is that the device's camera is not activated when a voice call is made. When making an audio call, ensure that the microphone button in your video-chatting app is not muted and the video button is disabled – e.g. by selecting the **Stop Video** button.

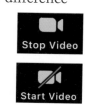

If the video button has a line through it, the video feed is off. Tap or click on the video button to remove the line, which will then display the video feed.

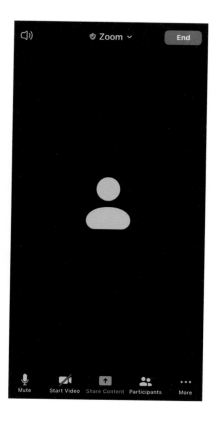

Group calls

Video-chatting apps are very sociable in that they enable large numbers of people to take part in a video chat at the same time. The maximum number varies between apps, but they can all accommodate calls for all but the largest families, or groups of friends.

Group calls can be customized to include all participants' video feeds at the same size (depending on the app).

The maximum number of people for a Zoom group video call is 100; for Skype it is also 100; for FaceTime it is 32; and for Messenger it is 50.

Alternatively, one video feed can be highlighted in the group chat.

...cont'd

Text chatting

A common feature with video-chatting apps is being able to conduct text chats, in a similar way to sending a text message on a smartphone or a tablet computer. Text chats can be done as a stand-alone feature – i.e. you just have a text conversation with someone and they can also be incorporated into a video or an audio call (i.e. while you are having a video chat, you can also send a text message).

Text chats are contained within their own window, whether they are conducted during a call or on their own.

Some video-chatting apps have a menu that can be used to add a selection of items to a text chat.

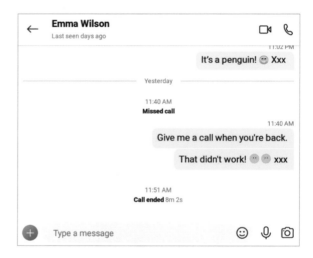

Additional content can be added to a text chat, including emoji icons and animated images.

Sharing

Some video-chatting apps have a function for sharing content from your device with other people in the call. This involves sharing what is on the screen of your device – e.g. if you have a photo or document that you want to share with the other people in the call, this can be done by activating the **Share** option and then opening the required item on your device. The other people in the call will then be able to see exactly what you have open on your device.

Content can be shared in a variety of ways, including screen sharing, sharing photos, and sharing any items that you have in an online storage service, such as iCloud or Dropbox.

Online storage services are ones that store your photos and documents in their own online computers (servers). This means that if something happens to your computer, you will still be able to access – and restore – your content from the online storage service. This is also known as cloud storage. Microsoft devices have access to the OneDrive online storage service; Apple devices have access to the iCloud online storage service; and Android devices have access to the Google Drive online storage service. Dropbox is another online storage service, and this is independent of the major technology companies.

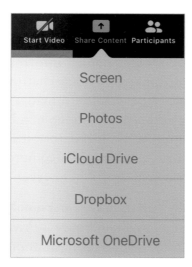

For screen sharing, a broadcast is created, displaying what you view on your own screen to the other participants.

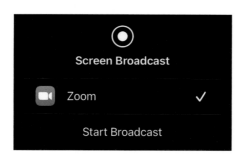

Security Issues

Video chatting should be a fun and enjoyable activity. To ensure that this is the case as much as possible, follow these points to stay as safe and secure as you can:

Some versions of video-chatting apps have to be downloaded from the app's own website, rather than from the device's linked app store. If this is the case, check on the website to see if there are any updates.

When you are in a video chat, someone could take a photo during the chat and post this to an online site, such as a social media site. This would then make the photo available to a wider audience, some of whom could look for any sensitive items in the background.

- **Download updates to your video-chatting apps whenever they are available**. On some devices they may be installed automatically (depending on how they are set up), and in other cases you may need to look in the app store from where the app was downloaded, to see if there are any updates.

- **Never accept a video or audio call from someone you do not know**. Scammers and cyber criminals are increasingly sophisticated and are turning to new ways of contacting potential victims. If you do not recognize a number that has called you, don't answer it.

- **Don't store too much personal information within a video-chatting app**. Most video-chatting apps allow you to store a range of personal information, such as email address, home address, and phone numbers. In general, if the information is not mandatory, it is best not to include it, just in case the app is compromised and the information becomes available to cyber criminals.

- **Check how the app uses your personal information**. Video-chatting apps should have a Privacy section (usually within the app's Settings options) that will tell you how the app and its parent company are using your personal information. It is worth taking some time to look through these details. In some cases, there are options for specifying how your information is used.

- **Check what is in the background of a video call when you are at home**. Even if video chats are just conducted with family and friends, it is important to check that your video feed does not have any sensitive items in the background, such as bank or credit cards. Also, any items that could cause any embarrassment should also be removed, if possible.

2 What You Need

This chapter looks at the equipment that is needed to get started with video chatting, from tablets to apps.

Devices

As with all forms of computing and digital communication, there is no shortage of devices that can be used for video chatting. These are the standard computing devices that we already use every day, for everything from browsing the web to sending emails: desktop computers, laptops, tablets and smartphones. In terms of video chatting, each type of device has its advantages, with tablets probably having the best overall combination of elements for video chatting.

Tablets

Tablet computers, such as Apple's iPad and the range of tablets that use the Android operating system, offer the best combination of screen size and mobility for undertaking video chatting in a variety of locations. The screens are generally large enough so that video calls can be viewed clearly with one or more participant(s), and they are mobile enough so that they can easily be taken to different rooms in the home, or to a range of locations when you are away from home and out and about.

Tablets also have all of the hardware elements required for video chatting (camera, microphone and speakers) built in to them, so it is just a question of obtaining the required apps and getting started. Ideally, a tablet with a stand should be used so that you do not have to hold it for the duration of a video call.

iPads use their own bespoke operating system, iPadOS, which cannot be used on other tablets. The Android operating system is used on a much wider range of tablets.

Smartphones

Smartphones offer the best option in terms of mobility, since most people generally have their phones with them most of the time. This means that video chatting can be undertaken on the move (with a mobile data connection – see page 27) and also with Wi-Fi in locations where this is provided, such as in coffee shops. In addition, a lot of smartphones are large enough so that the screen is effective for video chatting, particularly if it is one-to-one.

Some video-chatting apps allow for group chats involving dozens of people at one time. If you are going to be undertaking a large group chat, a smartphone screen may be too small to see everyone at an acceptable size.

Laptops

In terms of computing devices in the home, laptops have overtaken desktop computers for many people, since they are still powerful enough to perform everyday computing tasks, but they are easily portable around the

home and can also be taken away from the home, although they are not as portable as a tablet or a smartphone. In addition, laptops have large screens that are ideal for video chatting and can accommodate several family members in a call at the same time.

Desktop computers

Although sometimes now seen as the poor relation of the computing world, as it has been surpassed by more powerful but smaller and more compact devices, the humble desktop computer is still a viable option for video chatting, simply because it can be connected to a large monitor to make participants in a video chat appear much larger, and also facilitates chats with large groups of people, where they are all clearly visible.

Hot tip

Front-facing cameras are also known as selfie cameras, as they are used to take self-portraits (selfies) of the user. Look for a front-facing camera that has a minimum specification of HD (High Definition).

Hot tip

When using the camera for a video call, the video button in the video-chatting app is the one that shows whether the camera is activated or not. If there is a line through the icon, the camera is not activated. Ideally, the icon should not have a line through it, indicating that the camera is activated and the other participant(s) will be able to see you.

Cameras

Being able to see other people is a fundamental function of a video chat, and this is provided by a camera designed for this purpose. Most digital devices have built-in cameras that can be used for video chats, and they provide excellent functionality and quality. These types of cameras are located around the body of the device, usually in the middle at the top of the device. Smartphones and tablets generally have two cameras: a front-facing one and a back-facing one, which are on either side of the body of the device. The front-facing camera is the one above the screen of the device and this is the one that is used for video chatting, since you are looking at it when you look at the screen of your device.

If a digital device does not have a built-in camera, such as monitors for desktop computers, or if you are looking for increased quality from a camera for a video chat, an external one can be attached to your computing device. This is usually done with a USB cable connection, and the device is attached to your computer using some form of clip. This is a good option if you want to have a higher quality of video feed than the built-in camera on your device can provide: cameras using 4K UHD (Ultra High Definition) are some of the best on the market, at the time of printing.

Microphone and Speakers

Once a camera has been used to connect a video call and give a visual video feed, the next step is being able to hear what people are saying and to ensure that they can hear you. This is done with a microphone and a set of speakers. Most devices have these built in, so there is no absolute requirement to use separate ones. However, headsets or earphones are more effective for audio and video calls than the speakers and microphone on a digital device because they enable the two elements to be isolated and used separately, unlike the ones on a digital device, where only one element can be used at a time.

Headsets are the most sophisticated option for video chatting as they offer the best quality of sound and speech. These are connected to the digital device either with a USB cable or wirelessly, using a Bluetooth connection. This involves "pairing" the Bluetooth headset with the digital device, which is usually done in the

Settings section of the device. Pairing ensures that the headset and the device can communicate with each other.

Another audio option is using a pair of earphones or earbuds, such as those that are frequently provided with a smartphone or tablet. While these do not have the same sophistication as a dedicated headset, they still provide a good level of audio quality and are more portable than a

larger headset. As with headsets, these can be connected to a smartphone or tablet using a USB cable connection or a Bluetooth one.

For more details about getting connected for video chatting, see pages 32-35.

Wi-Fi Connection

Video-chatting calls are conducted over the internet, so one vital requirement is a suitable connection. This will be provided by your internet provider, which will usually be done through a broadband or fiber Wi-Fi connection.

Computing devices are required to link to a Wi-Fi router so that the device can then access the internet. Different devices will do this in slightly different ways, but in general the process is similar and is done through the device's **Settings** options. To set this up:

1. Access your device's **Settings** app and tap or click on it to open it

2. Access the **Wi-Fi** option, turn it **On** and select one of the available Wi-Fi networks

3. Enter the password for your Wi-Fi router, which will provide a connection to the Wi-Fi network

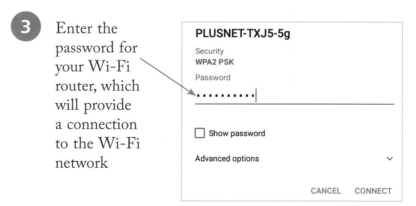

4. The connected Wi-Fi network is shown in the **Wi-Fi** settings

Mobile Connection

One increasingly popular way of conducting video chats is when people are away from home and on the move. This could be while traveling on public transport, or any other time when you are not at home. One option for doing this is by using a Wi-Fi hotspot, as shown on the previous page. However, for the ultimate flexibility in being able to conduct video chats, a mobile data connection (cellular data) can be used. This is usually done with a smartphone, but some tablets also have this functionality.

A mobile data connection is the one provided by your cell/mobile phone provider and it enables you to connect to the internet, and its related services, without the need for a Wi-Fi connection. The data plan that is available for your phone will depend on your mobile data provider: some companies offer unlimited data while others provide a certain amount of free data, over which you have to pay a charge. Most cellular data is used for accessing the internet, with functions such as streaming music or movies, or making video calls, being particularly data-intensive.

Details for cellular data can be accessed from the **Settings** options on your smartphone or compatible tablet. To do this:

Mobile data networks are denoted by the letter **G** on the top toolbar of digital devices. This stands for Generation and relates to the version of the network being used – e.g. 3G, 4G or 5G.

27

1 Access your device's **Settings** app and tap or click on it to open it

2 Tap or click on the **Cellular** option

3 Apply the Cellular settings as required. The **Cellular Data** option has to be **On** in order to access the internet using cellular data

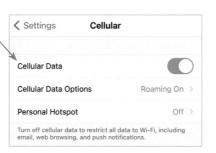

If you have a limit on your data plan, try to ensure that you do not go over it as you will incur additional expense. Video calls can use up data quickly, so try to restrict them if you are on a limited data plan.

Video-Chatting Apps

Once all of the hardware is in place for video chatting, calls can start to be made, using apps that are designed for this purpose. Some of the most popular video-chatting apps are Zoom, Skype, FaceTime and Messenger.

Zoom

Before the Covid-19 pandemic, relatively few people had heard of the Zoom app. However, with the requirement for people to stay at home much more during the pandemic, Zoom has quickly become established as one of the most popular video-chatting apps. Zoom is a multi-platform app that deals equally well with one-to-one chats, group chats and activities such as family quiz nights.

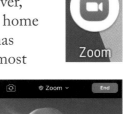

The Zoom interface is clear and easy to use, with options for starting video chats, joining them, inviting people to a chat, scheduling chats for specific times and dates, and even adding your own backgrounds for video chats.

Skype

Skype is one of the most enduring video-chatting apps, and it has been widely used for video and audio calls since it was first introduced in 2003. Now owned by Microsoft, it is a versatile and powerful communications tool that has a range of features in addition to being able to make video calls. It can be used on all major devices and operating systems.

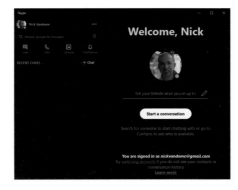

Don't forget

Zoom is also used as a business tool, so the terminology used for video chats is "Meetings".

FaceTime

For users of Apple devices – including iPhones, iPads, iMacs, Mac Minis and MacBooks – FaceTime is a logical choice for video chatting, as it is pre-installed on these devices. FaceTime is ideal for video chats with other Apple users, but its main drawback is that it is not compatible with Android mobile devices or Windows on desktop computers and laptops. If you need to make video calls from an Apple device to a non-Apple one, then a multi-platform video-chatting app such as Zoom, Skype, or Messenger is a better option, since it will work on all devices.

Video-chatting apps are free to download, and this can be done from the Android Play Store or the Apple App Store for mobile devices. FaceTime is pre-installed on Apple devices and Skype is pre-installed on some Windows devices. If not, it can be downloaded from the Microsoft Store.

29

Messenger

Messenger is a communication app owned by Facebook. The app was originally part of the Facebook interface, but it is now a stand-alone item. It contains excellent functionality for making and receiving video calls and it also has a range of options for sending text messages with a variety of content, either during a video call or as a separate text message.

Most video-chatting apps also have options for having text conversations.

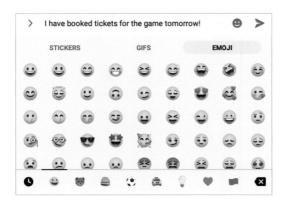

...cont'd

More video-chatting apps

With an increase in the number of people undertaking video chats on digital devices, it is only natural that there has also been a significant increase in the number of apps that can perform this function. Some of them are:

- **WhatsApp**. One of the most popular text-chat apps, WhatsApp also has a feature for video chatting. WhatsApp also works with Facebook Portal, which is a touchscreen device designed specifically for video calls between Messenger and WhatsApp users. See Chapter 9 for more details.

- **Snapchat**. Primarily used by the younger generation and teenagers, Snapchat can be used for video calls too.

- **Instagram**. A very popular app for sharing photos, Instagram has a chat feature that can quickly be converted for video calls.

- **Google Duo**. This is Google's main app for video calls and can be used on all major devices and operating systems.

- **Houseparty**. A relative newcomer on the video-chatting scene, Houseparty is a great option for bringing groups of people together.

- **Marco Polo**. Another newer video-chatting app, Marco Polo also provides a function for recording videos so that family and friends can view them at a time that suits them.

- **Microsoft Teams**. Although primarily a business tool, Microsoft Teams can be downloaded for free and has a powerful video-chatting option, in addition to a range of other useful functionality.

3 Getting Started

This chapter shows how to get up and running with everything you need for video chatting.

Getting Connected

Getting connected to the internet is an important step in being able to start video chatting. This is because video-chatting apps use the internet to link two devices in a video chat. Different devices and operating systems have the same general method for connecting to the internet – via a Wi-Fi router – although they all have their own steps.

Connecting with iOS and iPadOS

Apple's mobile devices, the iPhone and the iPad, use the iOS and iPadOS operating systems respectively and have the same process for connecting to the internet:

For iOS and iPadOS devices, a button in the Settings app is **On** when the button next to it displays green, as in Step 3.

1 Tap on the **Settings** app

2 Tap on the **Wi-Fi** option

3 Ensure the **Wi-Fi** button is in the **On** position

Wi-Fi	⬤

4 Available networks are shown here. Tap on one to select it

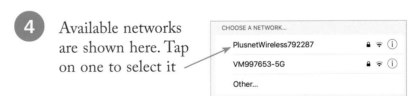

CHOOSE A NETWORK...

PlusnetWireless792287 🔒 📶 ⓘ

VM997653-5G 🔒 📶 ⓘ

Other...

5 Enter the password for your Wi-Fi router

Enter the password for "PlusnetWireless792287"

Cancel **Enter Password** Join

Password ●●●●●●●●●|

6 Tap on the **Join** button **Join**

7 Once a network has been joined, a tick appears next to it. This now provides access to the internet

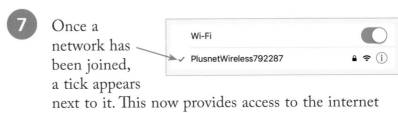

Wi-Fi ⬤

✓ PlusnetWireless792287 🔒 📶 ⓘ

Connecting with Android

Smartphones and tablets using the Android operating system can connect to the internet as follows:

1 Access your device's **Settings** app and tap on it to open it

2 Access the **Wi-Fi** option, turn it **On** and select one of the available Wi-Fi networks

3 Enter the password for your Wi-Fi router, which will provide a connection to the Wi-Fi network

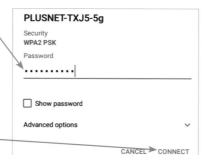

4 Tap on the **CONNECT** button

5 The connected Wi-Fi network is shown in the **Wi-Fi** settings

The Android operating system can be customized, to a certain degree, by different manufacturers of Android smartphones and tablets. This means that the Wi-Fi settings could differ slightly between different devices, but it is one area where the process is more or less standardized between manufacturers.

...cont'd

Connecting with Windows

Desktop computers or laptops using Windows 10 can connect to the internet as follows:

1 Click on the **Settings** app in the Start menu (or on the Taskbar if you have pinned it here)

2 Click on the **Network & Internet** option

3 Click on the **Wi-Fi** option in the left-hand sidebar

4 Ensure the **Wi-Fi** button is in the **On** position

5 Click on the **Show available networks** option

6 Click on the required option for providing Wi-Fi

7 Click on the **Connect** button

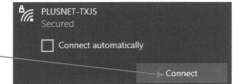

8 Enter the password for the Wi-Fi router and click on the **Next** button to connect to the Wi-Fi network

When a Wi-Fi network has been connected, it is displayed below the Wi-Fi heading, as in Step 4.

Connecting with macOS

Mac computers, including the iMac, Mac mini and MacBook, can connect to the internet as follows:

1 Click on the **System Preferences** icon on the Dock at the bottom of the screen

The operating system for Mac computers is called macOS.

2 Click on the **Network** button

3 Click on the **Turn Wi-Fi On** button

The **Turn Wi-Fi On** button is located in the top right-hand corner of the **Network** window.

4 Click in the **Network Name:** box

5 Click on the required network

6 Enter the password for the Wi-Fi router and click on the **Join** button to connect to the Wi-Fi network

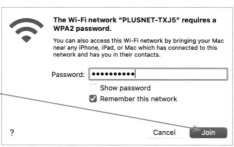

Mac computers can use the built-in FaceTime app for video chatting, and also other video-chatting apps such as Zoom, Skype, or Messenger, by downloading the relevant apps from the App Store or the app's own website.

7 The connected Wi-Fi network is shown in the **Network Name:** box

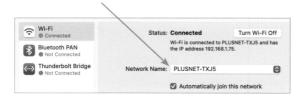

Finding Contacts

You cannot have a video chat if you cannot connect to anyone, so it is important to be able to add and access contacts with your video-chatting app.

Contacts with Zoom
To access contacts with the Zoom app:

For more details about finding contacts in Zoom and making calls to them, see pages 64-65.

36

1 Tap on the **Zoom** app

2 Tap on the **New Meeting** option

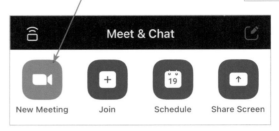

3 Tap on the **Invite** button in the bottom left-hand corner of the window **Invite**

4 Click on one of the options for contacting someone to join the video chat

Contacts with FaceTime

To access contacts with the FaceTime app:

1 Tap on the **FaceTime** app

2 Tap on this button to find a contact

3 Enter the details of a contact in the **To:** box, or click on this button

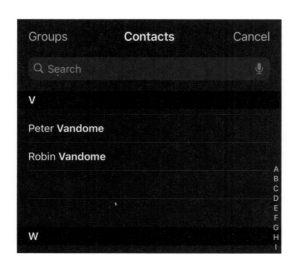

4 Contacts from your iPhone/iPad address book are shown. Tap on one to select someone for a chat

Hot tip

All participants in FaceTime calls require an Apple ID, which can be created when an Apple device is first set up. It can also be created on the Apple website at **https:// appleid.apple.com**

37

Hot tip

Use the **Search** box in Step 4 to search for specific people within your contacts.

Hot tip

For more details about finding contacts in FaceTime and making calls to them, see pages 132-133.

...cont'd

Contacts with Skype

To access contacts with the Skype app:

Don't forget

The illustrations on this page are from the Windows 10 version of Skype, rather than the mobile version.

Hot tip

For more details about finding contacts in Skype and making calls to them, see pages 108-110.

1 Click on the **Skype** app

2 Click on the **Contacts** button in the main window

3 Click on the **+ Contact** button

4 Enter the name of someone in the **Find people** box to contact people who are already registered on Skype, or click on **Invite to Skype** or **Add a phone number** to add contacts this way

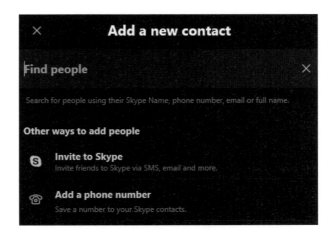

Contacts with Messenger

To access contacts with the Messenger app:

1 Tap on the **Messenger** app

2 Tap on the **People** button on the bottom toolbar

3 Tap on these buttons on the top toolbar to access the two areas for accessing contacts

4 Tap on this button to view all of your existing contacts, which are your Facebook contacts

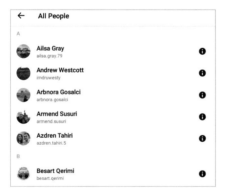

5 Tap on this button to add more contacts, including suggestions from Facebook

Accessing Settings

Customizing video-chatting apps is an important part in getting the most out of them, and this is done with Settings.

Settings for Zoom

To access the Settings options in the Zoom app:

For more details about using the Settings options in Zoom, see pages 60-63.

1 Open the **Zoom** app and tap on the **Settings** button on the bottom toolbar

2 Make selections for the way that Zoom operates, as required

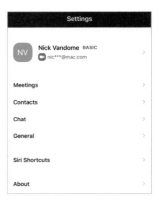

FaceTime differs from other video-chatting apps in that the settings are accessed from the Settings app on an iPhone or an iPad, rather from within the app itself.

Settings for FaceTime

To access the Settings options for the FaceTime app:

1 Tap on the **Settings** app on the Home screen of the device

2 Tap on the **FaceTime** option

3 Make selections for the way that FaceTime operates, as required

For more details about using the Settings options in FaceTime, see pages 130-131.

Settings for Skype

To access the Settings options in the Skype app:

1 Open the Skype app and click on the menu button on the main Homepage

The illustrations on this page are from the Windows 10 version of Skype, rather than the mobile version.

2 Click on the **Settings** button

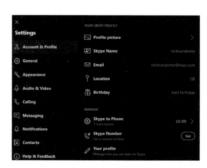

3 Make selections for the way that Skype operates, as required

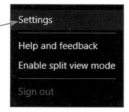

For more details about using the Settings options in Skype, see pages 102-106.

Settings for Messenger

To access the Settings options in the Messenger app:

1 Tap on your own account icon on the Homepage to access the full range of settings

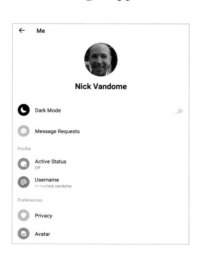

For more details about using the Settings options in Messenger, see pages 152-153.

41

Using Video

Video-chatting apps have a facility for voice-only calls, without video. This can be useful, particularly if you are on the move and want to use less data on your smartphone (since video uses more data than just audio). However, being able to see people is a fundamental part of video chatting, and it is a simple process to ensure that your video is on so that other participants can see you. To do this:

Hot tip

In some apps, the wording below the video icon describes what happens when the icon is clicked on or tapped, rather than its current state.

Stop Video

1 Start a video chat, using your preferred video-chatting app

2 If your video feed is not displayed, click or tap on the screen to access the app's toolbars

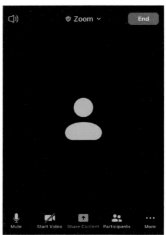

3 The video button will be displayed on a toolbar on the main screen. Click or tap on it so that there is no line through it. This will ensure that your video feed is visible

Hiding video

If you want to hide your own video feed during a video chat:

1 During a video chat in which your video feed is showing, click or tap on the screen to access the app's toolbars

Always check the state of your video feed when you first start a video chat, to ensure that it is on or off, as required.

2 Click or tap on the video button so that there is a line through it

3 Once this has been done, you will not be able to see your video feed, and neither will anyone else in the chat

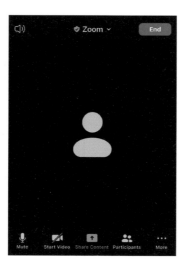

Using Audio

Controlling your microphone and being able to mute it and unmute it is an important function when in a video chat. The two main reasons for this are:

- **So the other participants can hear you when you start a video chat**: Numerous video chats have begun with someone saying, "I can't hear you, you'll have to unmute your microphone".

- **To ensure that group video calls do not descend into chaos, with everyone trying to talk at the same time**: If everyone except the person talking has muted their microphone then people can take it in turns to speak, with only one person unmuted at a time.

To mute and unmute your microphone during a video chat:

In some video-chatting apps, the microphone button also changes color, depending on whether it is muted or unmuted, in addition to having a line through it.

1 During a video chat, click or tap on the screen to access the app's toolbars

2 Click or tap on the microphone icon to change its state: if there is a line through it, it is muted; if there is no line through it, it is unmuted

3 In some video-chatting apps, if your microphone is muted, the other participants in the video chat will be able to see this with this icon next to your video feed

4 Following Good Practice

This chapter looks at some of the ways in which you can make your video chatting as effective and enjoyable as possible for everyone.

Hot tip

A number of video-chatting apps also have an option for blurring the background, to give more prominence to the person in the video chat – see page 49 for details.

Using Backgrounds

Video-chatting apps are becoming increasingly sophisticated in terms of their additional features, above and beyond providing video feeds. One of these features is the use of backgrounds that can be superimposed behind the participant in a video chat, to give a range of artistic, eye-catching and – at times – fantastical backgrounds.

Adding backgrounds
To add preset backgrounds to a video chat:

1 Access the app's **Background** section. In some apps this will contain **Filters** too

> Background and Filters

2 Click or tap on one of the images to add it as a background

3 The background appears behind your own video feed

... cont'd

Adding your own photos as backgrounds

In addition to using the preset backgrounds in a video-chatting app, it is also possible in some instances to use your own photos as the background for a video chat. This can help personalize your video chats with family and friends. To add your own photos as a background:

1 Access the app's **Background and Filters** section, as shown on the previous page, and click or tap on the add (**+**) button

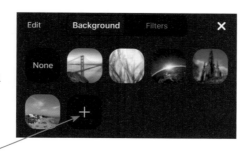

2 Navigate to the required photo from within your device's Photos app. Click or tap on it to select it, and click or tap on the **Done** button

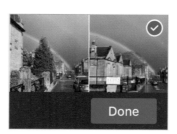

3 The selected photo is added as the background in the video-chat window

FaceTime does not have a function for adding backgrounds.

Once you have selected one of your own photos as a background, it becomes available in the **Background** tray of images.

...cont'd

Adding filters

Filter effects are another way in which the background to a video chat can be enhanced.

Use filter effects carefully, as they can become annoying for other people in a video chat if they are too elaborate or overpowering.

1 Access the app's **Background and Filters** section and click or tap on the **Filters** tab

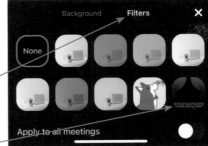

2 Click or tap on a filter to add that effect to your video feed

3 Filters offer a range of color and graphical effects that can be used during a video chat

Blurring the Background

To give yourself more prominence in a video chat, some apps have a facility for blurring the background. To do this:

1 Open your video feed. If there is a background that looks rather cluttered, this can be distracting for other people in the video chat

Backgrounds can be blurred before a video chat takes place or during the chat itself, if this functionality is available in the app.

2 Access the app's option for blurring the background and turn the button **On**

Enable background blur

3 The background is blurred, making it less distracting and giving more prominence to the subject in the video-chat window

Learning the Controls

All video-chatting apps have their own range of icons and controls. However, the main controls and their icons are consistent across most video-chatting apps. These include:

- **Microphone**. This is used to mute or unmute your audio feed. If it is muted, other people in the video chat will not be able to hear you.

- **Camera**. This can be used to display or hide your video feed. If it is off (with a line through it), other people in the video chat will not be able to see you.

- **Switch cameras**. This can be used to switch between the front-facing and back-facing cameras on a smartphone or a tablet. The front-facing camera is usually the one used to display yourself in a video chat.

- **Participants**. This can be used to invite more people to a video chat. Invites are generally sent by email or messaging apps.

- **Speakers**. This can be used to mute the speakers on your device, which means you will not hear anyone in the video chat.

- **Share content**. This can be used to share content during a video chat (such as photos) via email or messaging apps. It is also possible to share the screen from your device.

- **Menus**. This can be used to access more options for the video-chatting app being used, including the chat function during a call, adding emojis and adding backgrounds.

- **End call**. This is used to either leave a video chat while it is still in progress or end it for everyone in the chat.

Hot tip

Some menu options include the **Raise Hand** function. This can be used to raise a virtual hand in a video chat (which appears on your video feed) to indicate that you want to speak.

Setting Audio Levels

Getting the right level of sound for listening to other people in a video chat is an essential part of the process, to ensure that it is loud enough to hear other people but not too loud to impair your hearing. There are generally three ways in which the audio level can be set:

- **Using your device's volume controls**. These are usually situated on the body of the device for mobile devices, and a slider displays the volume levels.

- **Using your device's volume settings**. These are usually found in the **Sound** settings, and can be used to set the volume levels for incoming items.

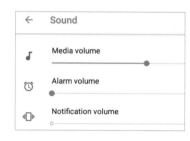

Don't forget

In some cases, it may take a combination of the volume options here to get the ideal setup for video chats.

51

- **Using the video-chatting app's volume controls**. Some video-chatting apps can use their own speaker options. If they have this, tap or click on the speaker icon during a video chat and select the **Speaker** option for the app. Drag the slider to set the volume, as required.

Beware

When using headphones or earbuds, it is important to keep the volume at a suitable level, which is usually slightly less than without headphones or earbuds, since the volume is closer to the ears with these options.

Understanding Audio Delay

One of the problems with video chats is knowing when to speak, to avoid talking over other people. This can be hard enough in real-life group conversations, but it is made even harder in the virtual world due to the inevitable impact of technology. Two areas to pay attention to in this regard are:

- **Wi-Fi connections**. A slow or intermittent Wi-Fi signal can have a significant impact on a video call, particularly in relation to when you hear someone speak. This can result in you hearing someone speak considerably later than they actually have done. Even a gap of a couple of seconds can greatly detract from a smooth audio conversation. If you are going to be using video chatting frequently, try to use the fastest Wi-Fi connection that you can, and check that the signal is reaching your device effectively. This can usually be done in the Wi-Fi settings for your device, most often with an icon showing the strength of the signal.

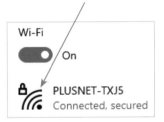

- **Using a device's speaker and microphone separately**. If you use your device's built-in speakers and microphone they cannot both operate at the same time, leading to possible audio delay when you are speaking or listening.  To solve this, use earbuds or headphones with a separate microphone that is designed for video calls.

Being Aware of Your Location

The increasing popularity of video chatting, and the ease with which it can be set up and used, means that it is possible to use it almost anywhere. However, this does not mean that you should. Some points to bear in mind when using video chatting (or not using it):

- **Be considerate of people around you**. Always think twice before having a video chat in a busy location, such as a restaurant or a coffee shop, or on public transport. People generally tend to talk more loudly when they are on a video (or audio) call, and this can become irritating for people near to you. This can then be exaggerated if several people are making calls, particularly if they are not using headphones or earbuds, with the result that both sides of the conversation can be heard. Try to think of other people and restrict your video calls to when you are at home or in a location where there are limited numbers of other people.

- **Be considerate of the other people on the call**. This is similar to considering your surroundings, but in relation to whoever else is on the call. It can be frustrating to be on a video call if there is a lot of background noise, particularly if it results in the other person having to talk more loudly.

- **Be aware about inadvertently giving out sensitive information**. If you are making a video call in a public place, anything you say could be overheard by someone else. Ensure that you do not mention any sensitive financial information or family details, such as someone's birthday, as this is one item that can be used by cyber criminals to build up a profile in order to steal a person's identity.

- **In general, the best option for video calls is in your own home**: you can make yourself comfortable and ensure that the call is as relaxed and as enjoyable as possible for everyone concerned.

Hot tip

Wireless earbuds are a good option if you are conducting a video call away from home. They connect to a smartphone or a tablet via Bluetooth technology and ensure that people near to you do not hear the other person in a call.

Managing Group Chats

There are a number of pitfalls to try to avoid when taking part in group video chats. Some things to consider are:

- Try not to talk over other people during a video chat. This can be harder to achieve than during a face-to-face chat, partly due to the reduced amount of body language available. It can be particularly problematic if everyone in the video chat is unmuted and trying to talk at the same time. A good way to overcome this is for the organizer of the video chat to mute all of the participants and then unmute them, as required. This can be done by the organizer asking individual people to speak, or by people using the Raise Hand function, if there is one.

- If you have something particular that you want to say, flag this up at the beginning of the group chat, or send a text message before the chat starts so that everyone else knows that you want to pass on some news.

- Never use derogatory or offensive language: if it's not acceptable face to face, it's not acceptable in an online video chat either.

- Don't shout at other people. It's not something that should be done in a face-to-face chat, and it can seem exaggerated in an online environment.

- Never say anything that you would not be happy to have played back to you. Even if a video chat is not being recorded, someone could still film it on a smartphone and so have a permanent record of it.

- If you are taking part in a video chat, be aware of your background and ensure that it is not distracting and does not contain any inappropriate items.

- Check your audio and video settings before the video chat starts, paying particular attention to your microphone to ensure it is muted, or unmuted, as required.

Beware

Group video chats can be harder to manage than face-to-face ones, but if you get frustrated about something, either mute your own microphone or let the other people know that you will be leaving the chat. This is better than saying something that you may regret later.

5 Zoom

In a relatively short period of time, the Zoom app has become one of the most popular apps for video chatting with family and friends. This chapter takes a detailed look at using Zoom, from obtaining the app to using it for effective video calls, text chats and sharing content.

Obtaining Zoom

Zoom has quickly become one of the most popular video-chatting apps and it is easy to see why: it is a multi-platform app that can be used on mobile devices, and Windows and Mac desktops and laptops, and it has a clear interface that enables people to spend more time on video chats, and less on working out the controls.

Zoom can be downloaded for iPhones using iOS, iPads using iPadOS, Android smartphones and tablets, Windows 10 desktops and laptops, and Mac desktops and laptops using macOS.

Zoom for iPhones and iPads

To download Zoom for iPhones and iPads, the process is the same for both:

1 Tap on the **App Store** app

2 Tap on the **Search** icon on the bottom toolbar

3 Enter "zoom" into the **Search** box

4 Tap on the **GET** button to download the Zoom app

5 Tap on the **Zoom** app to open it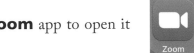

...cont'd

Zoom for Android smartphones and tablets
To download Zoom for smartphones and tablets using the Android operating system:

1 Tap on the **Play Store** app

2 Tap on the **Apps** button on the bottom toolbar

3 Tap in the **Search** box at the top of the window

4 Enter "zoom" into the **Search** box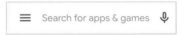

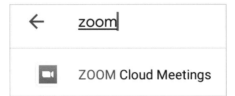

5 Tap on the **Install** button to download the Zoom app

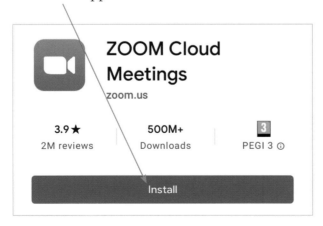

6 Tap on the **Zoom** app to open it

Tap on the **Enter** key on the keyboard to search for the exact word in the Search box in Step 4. If options are shown below the search word, tap on one to access it.

...cont'd

Zoom for Windows 10 devices

To download Zoom for Windows 10 desktop or laptop computers, the Zoom website is used, rather than downloading the app from an app store. To do this:

1 Access the Zoom website at **zoom.us**

2 Scroll down the Homepage and click on the **Meetings Client** link below the **Download** heading

3 Click on the **Download** button

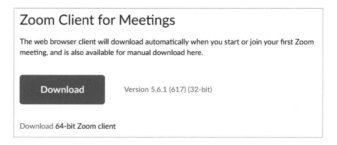

For some browsers, installation files for Windows that are downloaded from a website are available from a button in the browser. The installation files are also available in the **Downloads** folder in File Explorer.

4 Open the installation file from your browser or the **Downloads** folder, and click on the **Save File** button to start the installation process

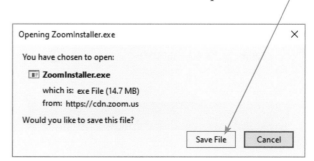

5 Open the Start menu and click on the **Zoom** app to open it

Zoom for macOS devices

As with the Windows version of Zoom, the macOS version that is used on Mac desktops and laptops (iMac, Mac mini, Mac Pro and MacBook) is downloaded from the Zoom website. To do this:

1 Access the Zoom website at **zoom.us**

2 Scroll down the Homepage and click on the **Meetings Client** link below the **Download** heading

Download

Meetings Client

3 Click on the **Download** button

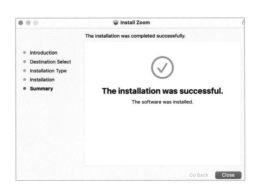

Zoom Client for Meetings

The web browser client will download automatically when you start or join your first Zoom meeting, and is also available for manual download here.

Download   Version 5.6.1 (560)

Or, for Macs with Apple Silicon chips, click here to download

Although the Download page in Step 3 looks the same as the one for downloading the Windows version of the app, the website recognizes that it is being accessed by a Mac computer and provides the appropriate file to be downloaded.

4 Open the installation file in the usual way for installing a new app. The installation process will proceed automatically, with a notification window when it is completed

The installation was completed successfully.

- Introduction
- Destination Select
- Installation Type
- Installation
- Summary

The installation was successful.

The software was installed.

Go Back   Close

To access the **Launcher** (displaying all of the apps on the device) on a Mac desktop or laptop, click on this button on the Dock at the bottom of the screen:

5 Open the **Launcher** and click on the **Zoom** app to open it

zoom.us

Zoom Settings

There is a range of settings for Zoom that can be used to customize it just the way that you want and get the most out of your video chats.

Accessing Settings

To access the Zoom settings:

Don't forget

The examples here are from the iPad version of Zoom. However, the functionality is the same for all other versions of Zoom. The terminology used in this chapter refers to tapping on a touchscreen device. For versions of the app on a desktop or laptop computer, the equivalent action will be clicking.

1 Tap on the **Zoom** app

2 The main interface is displayed

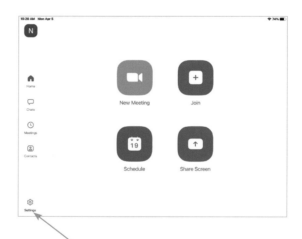

Hot tip

If the Zoom interface looks significantly different on your device than in the illustrations here, further help is available on our website. Visit **www.ineasysteps. com/products-page/ video-chatting-for- seniors-in-easy- steps** and click on **Resources**.

3 Tap on the **Settings** button

Settings

4 The main Settings categories are displayed. Tap on each category to view the options (see pages 61-63)

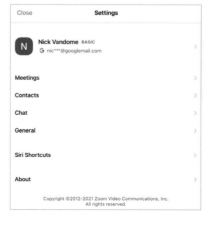

Meetings settings

The Meetings settings are the main ones for managing your video chats. To use them, tap on the **Meetings** button.

Meetings

1 The **Meetings** settings include options for: **Auto-Connect to Audio**, so that your audio is automatically **On**; **Always Mute My Microphone**, when joining a chat; and **Set Gallery View as Default**, for

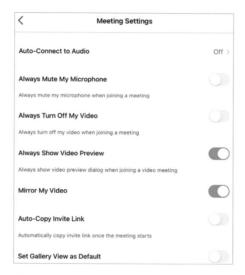

the format of the video windows during a video chat

2 Scroll down the window to view more of the Meetings settings. These include options for showing icons for people in the chat who are using audio only and not a

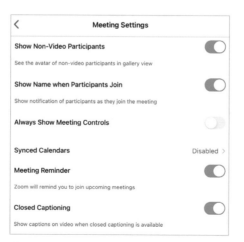

Meeting Reminder for scheduled meetings, and **Closed Captioning** for showing subtitles when someone is speaking in a video chat

The **Contacts** settings appear after the Meetings settings. These list contact requests that you have received. If you have not received any contact requests, the window will be empty.

Contacts

61

...cont'd

3 Scroll down the window to view more of the Meetings settings. These include options for showing the time of the video chat and also for touching up your appearance in your video feed

Chat settings

Text chats can be undertaken during a video chat and there are settings for this. The Chat settings are the main ones for managing your video chats. To use them, tap on the **Chat** button:

Chat

1 The **Chat** settings include options for displaying a link preview of text messages, keeping all unread messages at the top of the chat window, and managing unread messages

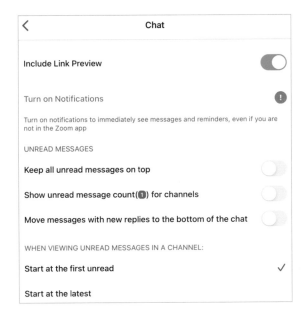

2 Scroll down the Chat settings window to view the **In-App Alert Sound** option, so that you are alerted when there is a new text message. Drag the button **On** to enable this

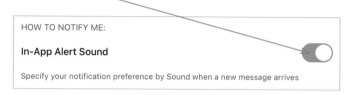

HOW TO NOTIFY ME:

In-App Alert Sound

Specify your notification preference by Sound when a new message arrives

Text-chat notifications can be specified to alert you of a new message, even when not using the Zoom app.

General settings

The General settings contain options that do not naturally fall into other categories. To use them, tap on the **General** button:

General

1 The **General** settings options include one for setting **Ringtones** when you are invited to a video chat or voice call

About settings

The About settings contain general information about Zoom. To use them, tap on the **About** button:

About

The **About** settings also have a **Report Problem** option to report an issue with the Zoom app.

1 The **About** settings include information about Zoom, such as the current version and its **Privacy Policy**

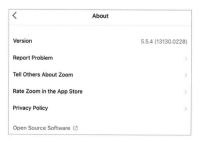

Adding Contacts

Video chats can be made in Zoom by starting a chat and then inviting people by email or a text message. However, contacts can also be added to the app so that you can quickly chat to your most popular contact. This can be done by inviting people so that you can add them to your Zoom app. To do this:

Hot tip

On smartphones, the **Contacts** button is on the toolbar at the bottom of the window.

1 Open the **Zoom** app and tap on the **Contacts** button in the left-hand sidebar

Contacts

2 Tap on the **Contacts** tab at the top of the Contacts window

3 Tap on the **+** button to start adding a new contact

4 Tap on the **Invite a Zoom Contact** option

Contacts   Channels   (+)

Invite a Zoom Contact

5 Enter an email address for the person you want to invite and tap on the **Add** button

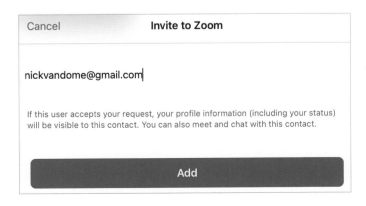

Cancel   **Invite to Zoom**

nickvandome@gmail.com

If this user accepts your request, your profile information (including your status) will be visible to this contact. You can also meet and chat with this contact.

Add

6 An invitation is sent to the named person. Tap on the **OK** button

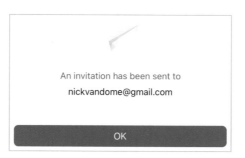

An invitation has been sent to

nickvandome@gmail.com

OK

When someone is invited as a contact they must have a Zoom account, or create one, in order to be able to accept the invitation.

7 The invited person has to accept the invitation, in which case they will be able to join the meeting

8 The details about the invited person are added in the **External Contacts** section of the Contacts window

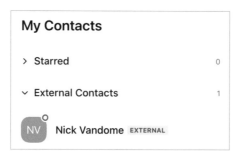

My Contacts

> Starred   0

∨ External Contacts   1

NV Nick Vandome EXTERNAL

9 Tap on a contact under the **My Contacts** section, to view their details in the right-hand panel and perform various functions within Zoom, such as starting a video chat (Meeting) or making a call

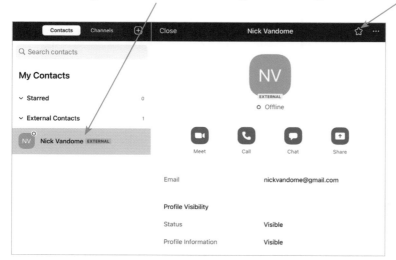

Tap on this button on the top toolbar when viewing a contact to add them as a **Starred** contact (favorite).

My Contacts

∨ Starred   1

NV Nick Vandome EXTERNAL

65

Starting a Zoom Call

Zoom calls can be started using contacts that have been added, as shown on pages 64-65, or people can be invited once a call has been started. To do this:

1 Open the **Zoom** app and tap on the **New Meeting** button

2 Select options for the meeting and tap on the **Start a Meeting** button

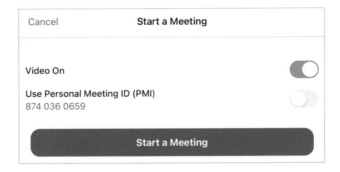

Depending on the version of Zoom being used, the toolbars may remain on the screen for the duration of a call – i.e. you do not have to tap on the screen every time you require the toolbars.

3 The Meeting (video chat) begins, with your own video feed displayed. Tap on the screen to access the top toolbar

4 Tap on the **Participants** button on the top toolbar to invite people

5 Tap on the **Invite** button in the Participants window

6 Select an option for inviting people to the video chat

7 For the **Send Email** option an email is created, with details of the meeting

If the **Copy Invite Link** option is used in Step 6, the copied link can then be pasted into another email or messaging app, rather than the default one used by the device.

Hot tip

67

8 For the **Send Message** option a text message is created, with details of the meeting

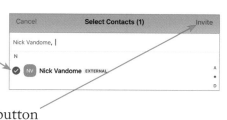

9 For the **Invite Contacts** option, tap on any contacts that have been added and tap on the **Invite** button

Scheduling a Zoom Call

Another option for starting a Zoom call is to schedule it for a specific date and time. It is also possible to create a recurring schedule, which is an excellent option if you have a video chat with family and friends at the same time every week/month. To schedule a Zoom call:

Hot tip

Numerous different calls can be scheduled, with their own dates and times.

1 Open the **Zoom** app and tap on the **Home** button

Home

2 Tap on the **Schedule** button

Schedule

3 The **Schedule Meeting** window contains options for scheduling a call

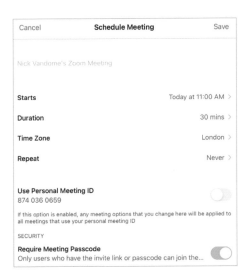

Cancel	Schedule Meeting	Save

Nick Vandome's Zoom Meeting

Starts — Today at 11:00 AM >
Duration — 30 mins >
Time Zone — London >
Repeat — Never >

Use Personal Meeting ID
874 036 0659

If this option is enabled, any meeting options that you change here will be applied to all meetings that use your personal meeting ID

SECURITY

Require Meeting Passcode
Only users who have the invite link or passcode can join the...

4 Tap in the top text box and enter a name for the chat

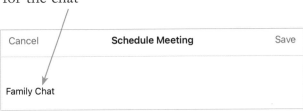

Cancel	Schedule Meeting	Save

Family Chat

5 Tap on the **Starts** button to set a date and a time for the start of a call

Starts

6 Drag the barrels to set a date and time for the start of a call. Tap on the **Done** button

Done

Thu Jul 22	9	15	
Fri Jul 23	10	30	
Sat Jul 24	11	45	AM
Sun Jul 25	**12**	**00**	**PM**
Mon Jul 26	1	15	
Tue Jul 27	2	30	
Wed Jul 28	3	45	

7 Tap on the **Duration** button   Duration

8 Drag the barrels to set the duration of a call. Tap on the **Done** button

Done

	15
	30
0 hours	**45 min**
1	0
2	15
3	30

The Date and Time barrels for the start of a meeting in Step 6 move independently from each other.

Zoom calls with more than three people are limited to 40 minutes with the free Basic plan version. If a meeting is set for longer than this, you can stop the meeting before 40 minutes and then start another one. If the time for a scheduled meeting is longer than 40 minutes, a box will appear alerting you to this. Tap on the **OK** button to remove the alert box.

> Your Zoom Basic plan has a 40-minute time limit on meetings with 3 or more participants
>
> OK

...cont'd

The **Every Year** option for **Repeat** is a good idea if you want to remember to call someone on their birthday every year.

The **Require Meeting Passcode** in Step 12 does not need to be used, particularly with family and friends, but it can give an extra level of security for people joining the call.

The **Enable Waiting Room** option in Step 12 can be used so that people can only join a call once the host admits them. If this option is **Off**, they will be part of the call as soon as they join it.

9 Tap on the **Repeat** button

Repeat

10 Select an option for repeating the meeting, as required, and tap on the **Done** button

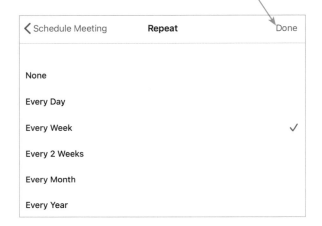

11 Swipe up the Schedule Meeting page and drag the buttons **On** or **Off** for **Host Video On** and **Participant Video On**, to control the video feeds when a meeting starts

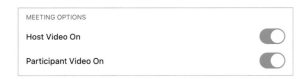

12 Swipe up the Schedule Meeting page and drag the buttons **On** or **Off** for **Require Meeting Passcode** and **Enable Waiting Room**, to set security options for the call

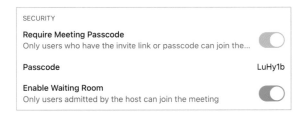

13 Swipe up the Schedule Meeting page and tap on the **Advanced Options** button

14 Make the selections for the Advanced options

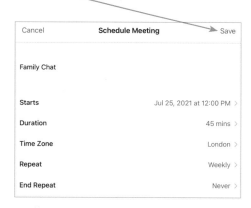

Allow Participants to Join Before Host

Automatically Record Meeting

Approve or Block Entry for Users from Specific Countries/Regions   None >

Don't forget

The Advanced options are for allowing participants to: join a meeting before the host; automatically record a meeting; and approving, or blocking, participants from specific countries or regions.

15 Once all of the options have been completed, review the details of the scheduled meeting, and tap on the **Save** button

Cancel | **Schedule Meeting** | Save

Family Chat

Starts   Jul 25, 2021 at 12:00 PM >

Duration   45 mins >

Time Zone   London >

Repeat   Weekly >

End Repeat   Never >

16 Once a scheduled meeting has been saved, an email is automatically generated for inviting people. Enter a recipient into the **To:** box and send the email

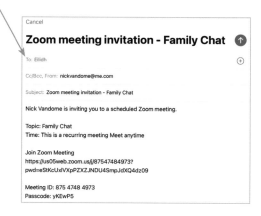

Cancel

Zoom meeting invitation - Family Chat ⬆

To: Eilidh   ⊕

Cc/Bcc, From: nickvandome@me.com

Subject: Zoom meeting invitation - Family Chat

Nick Vandome is inviting you to a scheduled Zoom meeting.

Topic: Family Chat
Time: This is a recurring meeting Meet anytime

Join Zoom Meeting
https://us05web.zoom.us/j/87547484973?
pwd=eStKcUxlVXpPZXZJNDU4SmpJdXQ4dz09

Meeting ID: 875 4748 4973
Passcode: yKEwP5

Hot tip

If a passcode has been set for the meeting, this will be included in the email in Step 16.

...cont'd

Accepting a scheduled meeting

If you receive an invitation for a scheduled Zoom meeting, it is possible to join the meeting and also add it to your calendar so that you have a record of a future meeting:

1 A Zoom meeting invitation email contains a link to the meeting, and any details required to join the

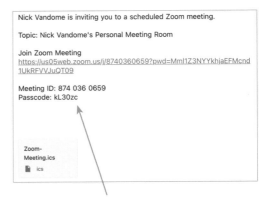

meeting, such as Meeting ID and Passcode. At the time of the meeting, click or tap on the link to join the meeting

2 At the bottom of the email, click or tap on the **ics** icon to add the scheduled meeting to your calendar

3 The scheduled meeting is added to the default calendar app on the device on which the **ics** file was opened

Hot tip

An **ics** file is an iCalendar one, which is a common standard that can create calendar entries for all of the main computing devices and operating systems.

Viewing scheduled meetings

Once scheduled meetings have been set up, it is possible to view them all and edit them, if required. To do this:

1 On the Zoom Homepage, tap on the **Meetings** button

2 All of the scheduled meetings are listed on the **Upcoming Meetings** page

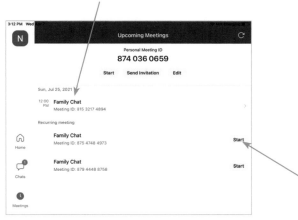

3 Tap on a meeting to view its details. Tap on the **Edit** button to edit the elements of the meeting

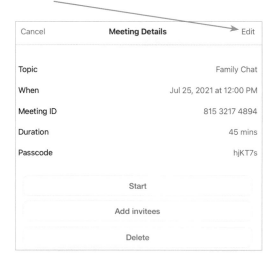

The **Upcoming Meetings** page contains details of single meetings and also recurring ones.

Scheduled meetings do not start automatically; you still have to manually start them by tapping on the **Start** button next to the meeting in Step 2.

Use the buttons at the bottom of the window in Step 3 to, from top to bottom: **Start** a meeting; **Add Invitees**; or **Delete** a scheduled meeting.

73

Joining a Zoom Call

There are two main ways of joining a Zoom call: when you have been invited via a link in an email or a text message or by using the **Join** option within the Zoom app.

Joining a call using a link

To join a Zoom call that is already in progress, when you have been invited with an email or a text message:

Don't forget

Although the Meeting ID and Passcode are shown in an email invitation, as shown in Step 1, they are not required to join a meeting in progress, if you use the link in the email to join.

1 Click or tap on the link in the email or text message

> **Please join Zoom meeting in progress**
>
> Join Zoom Meeting
> https://us04web.zoom.us/j/73397921403?
> pwd=ODJkSkViR0M4V0s3Zm9xTEdPeFBldz09
>
> Meeting ID: 733 9792 1403
> Passcode: x3HcPw

2 The Video Preview window displays your video feed as it will appear in the call. Tap on the **Join with Video** button

Video Preview

Always show video preview dialog when joining a video meeting

Join without Video   Join with Video

3 The host of the meeting has to admit you to the call

> Please wait, the meeting host will let you in soon
>
> Nick Vandome's Zoom Meeting

4 Once the host has selected the **Admit** button, you will be connected to the call

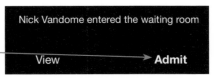

> Nick Vandome entered the waiting room
>
> View   **Admit**

Joining a call using the Join option

To join a Zoom call using the **Join** option in the Zoom app:

1 Open the **Zoom** app and tap on the **Home** button

Home

2 Tap on the **Join** button

Join

3 Enter the Meeting ID, and tap on the **Join** button

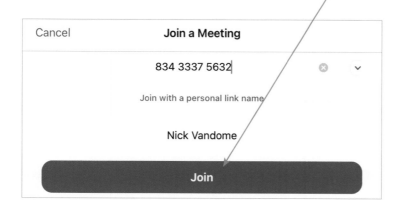

Cancel	Join a Meeting	
	834 3337 5632	
	Join with a personal link name	
	Nick Vandome	
	Join	

Hot tip

The Meeting ID is required in Step 3 to ensure that you join the correct meeting. The Meeting ID will be included in an email invite for the meeting.

75

4 Enter the passcode for the meeting, if required, and tap on the **Continue** button

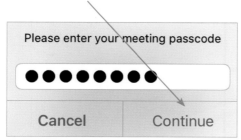
Please enter your meeting passcode
●●●●●●●●
Cancel   Continue

5 Join the meeting in the same way as from Step 2 on the previous page

During a Zoom Call

Once a Zoom call is up and running, there are a number of options for conducting and managing the call:

The thumbnail in Step 1 cannot be moved around the screen – i.e. you cannot press on the thumbnail and drag it to another position on the screen.

1 For a one-to-one call, your own video feed appears as a thumbnail in the bottom right-hand corner of the screen

2 Tap your own thumbnail video feed to swap it with the other participant's

3 Tap on the minus symbol (−) on the video thumbnail to hide it

4 Tap on the video icon to reinstate the video thumbnail

5 Tap on the plus symbol (+) on the video thumbnail to create thumbnails for both of the participants in the call (the other participant's video feed is still the main one on the screen)

Hot tip

Expanding the thumbnails as in Step 5 is particularly effective for group chats. See pages 81-82 for details.

77

6 Tap on this icon on the thumbnails at the bottom of the screen to return to a single-thumbnail format

...cont'd

Even if someone's video is not visible, you will still be able to hear them, as long as they have not muted themselves.

If nothing is actioned on the toolbars they will disappear after a few seconds, depending on the device being used.

 7 If the other participant in a one-to-one call turns off their camera or moves away from Zoom to use another app, their video feed will appear blank

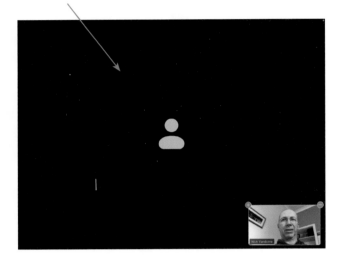

Using the toolbars in a call

During a call there are numerous options that can be accessed from the Zoom toolbars:

1 Tap on the screen to access the relevant toolbars

2   Tap on the **Switch to Gallery View** button

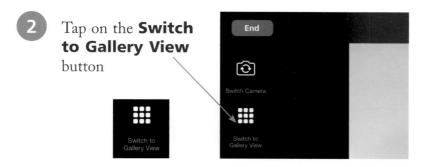

3   The video feeds appear at the same size together on the screen

Gallery View is particularly effective for group chats. See pages 81-82 for details.

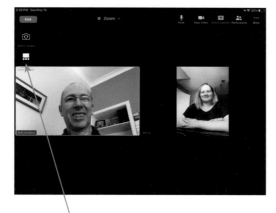

4   Tap on the **Switch to Active Speaker** button to return to the other participant's video feed appearing as the main one

...cont'd

The **Switch Camera** option is a good one for showing the other participants in a video chat your surroundings, particularly if they are notable or picturesque.

The function for switching cameras is only available on smartphones and tablets.

5 Tap on the **Switch Camera** button to reverse your own video feed

6 On the top toolbar, use these buttons, from left to right, to: mute your own sound; turn off your own video; share content from your device; and invite more participants to the call

7 Tap on the **More** menu button on the top toolbar

8 The **More** menu options are available for the current chat. Tap on an option to apply it, or access its further options

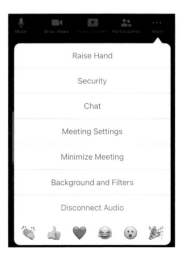

Managing a Group Call

Group chats can require a bit more management than one-to-one chats, simply because there are more video feeds involved. To manage a group call with more than two people:

1 By default, the call will open in active speaker mode, which displays the video feed of whoever is talking in the main video window

2 The other video feeds are shown as thumbnails at the bottom of the screen, including the active speaker

3 Tap on this button in the top right-hand corner of the thumbnail bar to hide it

The main video window in Step 1 will change when someone else speaks – i.e. they become the active speaker in the main window.

The active speaker thumbnail has a green border around it at the bottom of the screen.

...cont'd

4 If the thumbnail toolbar is hidden, only the active speaker is visible, and this changes when someone else speaks and becomes the active speaker

Tap on the **Switch to Active Speaker** button in Step 7 to return to the active speaker in the main window and the other video feeds as thumbnails at the bottom of the screen.

5 Tap on this button to restore the thumbnail bar

6 Tap on the **Switch to Gallery View** button to view all of the video feeds equally

7 Gallery View displays all of the video feeds at the same size in the main window. Whoever is the active speaker still has a green border around their window

Raising a Hand

During a Zoom call that has multiple participants, such as a family call or a quiz, the Raise Hand feature is a good option for determining who can speak, and when. This can reduce the chance of several people trying to talk at once. To use the Raise Hand option:

1 Access the **More** menu options, as shown in Step 7 on page 80, and tap on the **Raise Hand** option

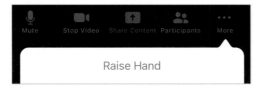

as shown in Step 7 on page 80

2 The hand icon is displayed on your own thumbnail video feed

3 The hand icon is displayed on the other participant's screen, with your name next to it

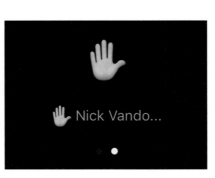

4 Tap on the **More** menu and tap on the **Lower Hand** option to lower your hand

Using Text Chat

Another option that is available with a lot of video-chatting apps is being able to send text messages during a video chat. This adds an extra option in terms of how you communicate with the app. To do this with Zoom:

1 Access the **More** menu options, as shown in Step 7 on page 80, and tap on the **Chat** option

2 The **Chat** window opens over the existing video-feed window

3 Tap here to select who will see your text messages: everyone in the call or specific people

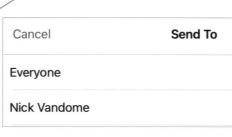

4 Enter a message in the text box at the bottom of the **Chat** window and tap on the **Send** button

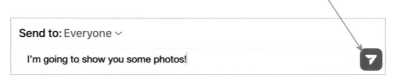

5 As more text messages are added, the conversation continues down the **Chat** window

6 If there are unread text messages, a red notification appears on the **More** menu

7 The number of unread messages also appears on the **Chat** button on the **More** menu. Tap on the **Chat** button to view the new messages

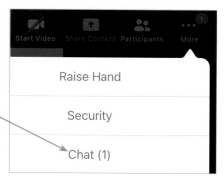

8 To stop the text-chat notifications from appearing on screen, tap on this button in top right-hand corner of the Chat window

9 Tap on the **Mute** button to stop notifications appearing for new text messages

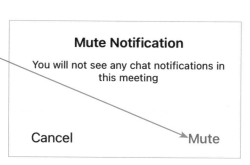

Hot tip

Text notifications appear as a banner on the video-feed window, even if the Chat window is not open. If notifications are muted, you have to open the Chat window to see any text messages.

Sharing Content

Zoom has various ways in which content can be shared with other participants during a video call.

Screen broadcasting

One way of sharing content in a Zoom call is screen broadcasting, which shows all of the participants in the call exactly what you are doing on your device. To do this:

Don't forget

The other **Share Content** options in Step 2 include sharing photos (see pages 88-89) and also for sharing content that you have in online storage and sharing apps, such as iCloud for Apple devices, Dropbox and Microsoft OneDrive.

1   Tap on the screen during a video call to access the toolbars, and tap on the **Share Content** button

2   Tap on the **Screen** option to start broadcasting all of your actions on your device

3   Tap on the **Share Broadcast** option to start sharing what you are doing on your device

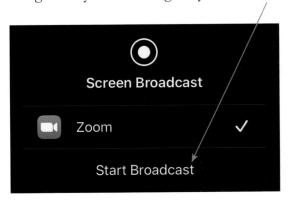

4 Screen broadcasting will show all of the participants in the call what is happening on your device. For instance, if you open a photo in the Photos app, then return to the Home screen and then open a web page, this is what the other participants in the call will see:

Beware

By default, screen broadcasting will display any notifications that you receive while you are broadcasting your screen. For instance, if you receive any email or text message notifications, these will be visible to everyone in the video call. To prevent this, activate the **Do Not Disturb** option on your device, or turn **Off** notifications for specific apps.

5 To stop screen broadcasting, tap on this icon at the top of the window

6 Tap on the **Stop** button to stop broadcasting your screen

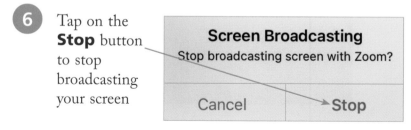

Screen Broadcasting
Stop broadcasting screen with Zoom?

| Cancel | Stop |

...cont'd

Sharing photos

Video chats with family and friends are a good opportunity to show people your photos, without the need to send them via email or a messaging app. To do this in Zoom:

If you select the **Don't Allow** option in Step 2, you will not be able to share any of your photos during a Zoom call.

1 Tap on the **Share Content** button, as shown on page 86, and tap on the **Photos** option

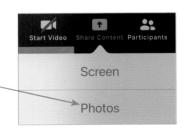

2 Select a permissions option for Zoom to access your photos

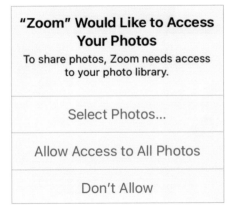

"Zoom" Would Like to Access Your Photos

To share photos, Zoom needs access to your photo library.

Select Photos...

Allow Access to All Photos

Don't Allow

3 For the **Select Photos...** or **Allow Access to All Photos** options, tap on the photo(s) that you want to share and tap on the **Done** button

4 The selected photo(s) is displayed on your screen and this is what other participants will see

5 Tap on this button to access the editing toolbar for the photo

To retain editing changes that you have made to a photo in Zoom, tap on the **Save** button on the editing toolbar in Step 5. This is saved as a new photo in your Photos app, and the original is unchanged.

6 Apply any editing changes, as required. These will also be seen by the other participants in the call

7 Tap on the **Stop Share** button on the top toolbar to stop sharing your screen or any content

Zoom Security

Security is a major issue in any computing environment, and Zoom has a range of options to ensure that you have as much control as possible over your calls. To access these:

1. Access the **More** menu options, as shown in Step 7 on page 80, and tap on the **Security** option

2. The **Security** options are listed in their own window. Use the buttons to turn them **On** or **Off**

3. Drag the **Lock Meeting** button **On** to prevent actions being carried out within the meeting

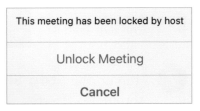

4. If an action is performed that is prevented by **Lock Meeting**, such as inviting new participants, this message is displayed

The Zoom security options are only available to the host of a meeting – i.e. the person who started the meeting.

5 Drag the **Waiting Room** button **On** to use this function
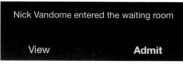
so that you can admit participants to a meeting, rather than them automatically joining

6 If **Waiting Room** is **On**, you will be alerted when someone wants to join a
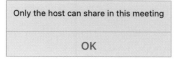
meeting. Tap on the **Admit** button to let them join

Make changes to the security settings before you invite anyone to a meeting you have started, to ensure that the changes will be applied to the active meeting.

7 Drag the **Share Screen** button **Off** to prevent other participants from sharing their screens

8 If **Share Screen** is **Off**, anyone trying to do this will be presented with this screen

Only the host can share in this meeting

OK

9 Tap on the **Chat with** option

Chat with   Everyone >

10 Select an option for who you would like participants to be able to send text messages to, via Zoom during the meeting

Allow Participants to Chat with

No One

Host Only

Everyone Publicly

Everyone

Ending a Zoom Call

Any participant in a Zoom call can leave the call. In addition, the host of a call (i.e. the person who started the call) can end the call for all participants.

Leaving a Zoom call

To leave a Zoom call that someone else has started:

Don't forget

Only the host of a call can end it for everyone.

1 During the call, tap on the **Leave** button at the left-hand side of the top toolbar

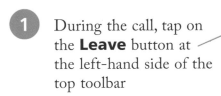

2 Tap on the **Leave Meeting** button to leave the meeting, or the **Cancel** button to continue

Ending a Zoom call

To end a Zoom call of which you are the host – i.e. you have started:

Beware

If you are the host of a call and you plan to end it for everyone, make sure that you tell all of the participants so that it does not come as a surprise.

1 During the call, tap on the **End** button at the left-hand side of the top toolbar

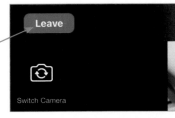

2 Tap on the **End Meeting for All** button to stop the call for everyone, or tap on the **Leave Meeting** button to just leave it yourself

6 Skype

Skype is a well-established video-chatting app that has an impressive range of features. This chapter shows how to get started with Skype, customize it for your own needs, and conduct a range of video and text chats.

Obtaining Skype

Skype is a versatile, multi-platform app that can be used on devices with all of the main operating systems.

Skype for iPhones and iPads

To download Skype for iPhones and iPad, the process is the same for both:

1 Tap on the **App Store** app

2 Access the Skype app and tap on the **GET** button to download it

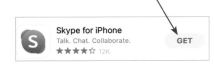

3 Tap on the **Skype** app to open it

Skype for Android smartphones and tablets

To download Skype for smartphones and tablets using the Android operating system:

1 Tap on the **Play Store** app

2 Access the Skype app and tap on the **Install** button to download it

3 Tap on the **Skype** app to open it

...cont'd

Skype for Windows 10 devices

Since Skype is now a Microsoft product, it comes pre-installed on some Windows 10 devices. However, it can also be downloaded from the Microsoft Store. To do this:

1 Click on the **Microsoft Store** button

2 Access the Skype app and click on the **Get** button to download it ⟶

3 Open the Start menu and click on the **Skype** app to open it

Skype for macOS devices

Unlike with the other versions of Skype, macOS devices do not have a version in the App Store. Instead, the app has to be downloaded from the Skype website. To do this:

1 Go to the website at **www.skype.com/en/get-skype** and click on the **Get Skype for Mac** button

2 Open the **Launcher** and click on the **Skype** app to open it

Don't forget

Versions of Skype for Windows desktops and laptops, and mobile devices, can also be downloaded from the Skype website.

Hot tip

To access the **Launcher** (displaying all of the apps on the device) on a Mac desktop or laptop, click on this button on the Dock at the bottom of the screen:

Setting Up Skype

When you first start using Skype, there are a few setup steps that have to be performed. To do this:

Skype was bought by Microsoft in 2011, which is why a Microsoft Account is required to use it.

1 Tap or click on the Skype icon to open the app. A Microsoft user account is required to use Skype. Tap or click on the **Sign in or create** button

A Microsoft Account username can be an email address.

2 If you already have a Microsoft Account, enter your username details and tap or click on the **Next** button. If you do not have an account, tap or click on the **Create one!** button

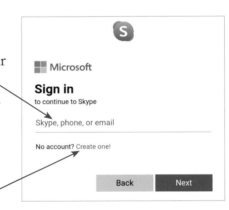

3 For an existing Microsoft Account, enter your password and tap or click on the **Sign in** button

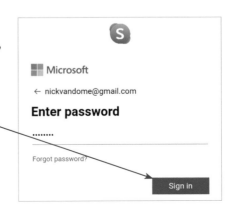

...cont'd

4 During the setup process, contacts can be imported from your address book. To do this, tap or click on the **Sync contacts** button, or to skip this step, tap or click on the **Skip** button

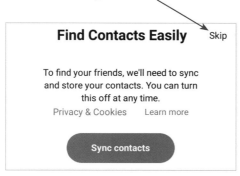

5 Tap or click on the **Allow** button to enable Skype to use the audio features on your device

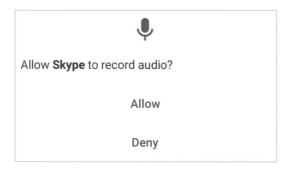

If you do not allow Skype access to the audio features on your device (microphone and speakers) and the camera, you will not be able to conduct video chats properly.

6 Tap or click on the **Allow** button to enable Skype to use the camera on your device

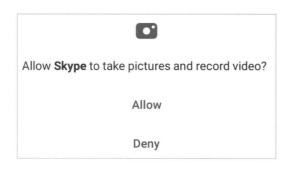

Skype Interface

The mobile and desktop/laptop versions of Skype are different in terms of their interface, although the functionality is the same across different devices.

Skype for mobile devices

The mobile versions of Skype have the same interface for Android and iOS/iPadOS devices. To use them:

The **Chats** panel is like a log of the activity that you have undertaken on Skype.

Calls and video chats are free between Skype users over Wi-Fi. Calls can also be made to non-Skype numbers, in which case credit has to be added to your Skype account, by tapping on the **Get credit** button in Step 2.

Contacts can be added to your Skype account from your own address book on the device on which you are accessing Skype, and also by adding other Skype users. See pages 108-110 for more details about adding contacts.

1 Tap on the **Chats** button on the bottom toolbar. The **Chats** panel contains details of texts and calls that have been made

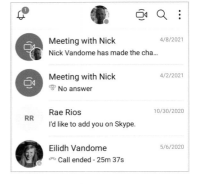

2 Tap on the **Calls** button on the bottom toolbar. The **Calls** panel contains details of video and voice calls that have been made

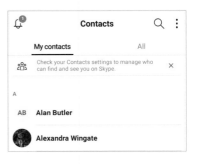

3 Tap on the **Contacts** button on the bottom toolbar. The **Contacts** panel contains details of people you have added to Skype, and options for adding more

98

Skype for Windows desktops and laptops

The interface for the version of Skype for Windows devices contains a larger window, but with similar controls. To use it:

1 Open the Skype app on a Windows desktop or laptop. The window contains the functionality in the left-hand panel and a main window area

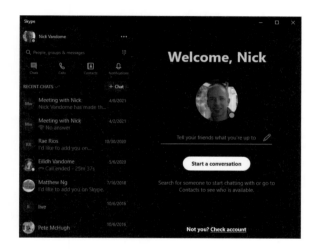

2 Click on the **Chats** button on the top toolbar

3 The **Chats** panel contains details of text messages and calls that have been made

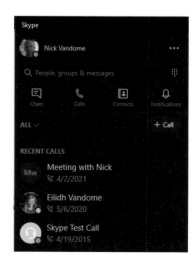

As you start using Skype, the main window displays details of your activity with the app. It also displays text chats.

...cont'd

Don't forget

Video chats open in the whole Skype interface – i.e. they take over both the left-hand panel and the main window.

Hot tip

Read receipts alert you to when someone has read a message that you have sent to them via Skype.

4 Click on the **Calls** button on the top toolbar. The **Calls** panel contains details of audio and video calls that you have made and received

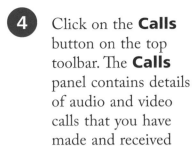

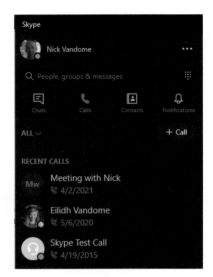

5 Click on the **Contacts** button on the top toolbar. The **Contacts** panel contains details of people you have added to Skype, and options for adding more

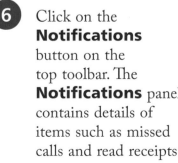

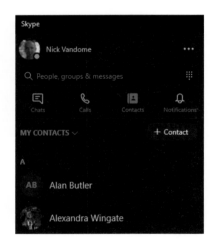

6 Click on the **Notifications** button on the top toolbar. The **Notifications** panel contains details of items such as missed calls and read receipts

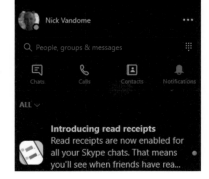

...cont'd

Skype for macOS desktops and laptops
The interface for the Skype app on Mac desktops and laptops using macOS is similar to the Windows version:

1 Open the Skype app on a macOS desktop or laptop

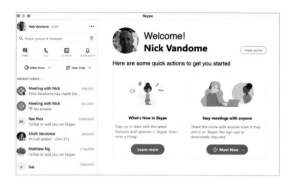

2 Use the buttons at the top of the left-hand panel to access the main functions: **Chats**, **Calls**, **Contacts** and **Notifications**

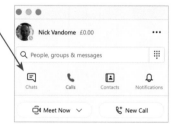

3 The main window displays details relating to the item selected in the left-hand panel

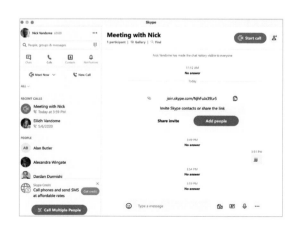

The overall appearance of the Skype interface on all devices can be set to Light or Dark mode, within the Skype settings. See page 104 for details.

Skype Settings

There is a range of settings for Skype that can be used to customize it just the way that you want and get the most out of your video chats, and more.

Accessing Settings

To access the Skype settings:

The examples in this chapter are from Skype on an Android tablet. The terminology used in this chapter refers to tapping on a touchscreen device. For versions of the app on a desktop or laptop computer, the equivalent action will be clicking.

1 Tap on the **Skype** app to open it

2 Tap on your own account icon, at the top of the screen

3 Tap on the **Settings** button

4 The main Settings categories are listed. Tap on each one to view its contents

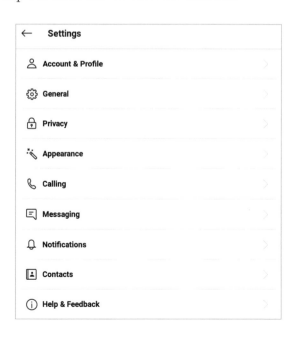

If the Skype interface looks significantly different on your device than in the illustrations here, further help is available on our website. Visit **www.ineasysteps. com/products-page/ video-chatting-for- seniors-in-easy- steps** and click on **Resources**.

5 Tap on the **Account and profile** setting to select options for: changing your profile picture; changing your Skype name; changing your email address linked to your Skype account; changing your location; and adding your birthday, if desired

← Account and profile	
YOUR SKYPE PROFILE	
🖼 Profile picture	>
💳 Skype Name	nickvandome
✉ Email	nickvandome@mac.com
📍 Location	GB
🎁 Birthday	Add birthday
MANAGE	
Ⓢ **Skype to Phone** 1 Subscription	£0.00 >
📞# **Skype Number** Get a second number	(Get)
✏ **Your profile** Manage how you are seen on Skype	
🗔 **Your account** Manage your payments & subscriptions	
✕ **Close your account** Delete your Microsoft account	>

Your Skype name is the one that appears to other Skype users when you call them or send them a text message.

103

6 The **MANAGE** section of the **Account and profile** setting has options for making Skype calls and for managing your profile

It is worth taking some time to look through the Skype settings as this will give you more confidence when using the app.

7 Tap on the **General** setting to select options for the default language used on Skype and for settings for translating different languages

← General	
GENERAL	
A🈁 Language	
TRANSLATION	
Translation settings	

...cont'd

Cookies are small programs that gather information about the user of an app or a website. They are useful for helping an app remember certain actions that have been performed by a user, but it is important to be aware of how cookies are used by apps and websites.

Light or **Dark** mode is applied to the whole of the Skype interface. Dark mode can make items clearer, but it comes down to a matter of personal preference as to which one people prefer.

8 Tap on the **Privacy** setting to select options including sharing your location and restricting calls from only your contacts, and for information about how cookies are used on Skype

9 Tap on the **Appearance** setting to select options for the look and feel of the Skype interface, including the color and using **Light** or **Dark** mode

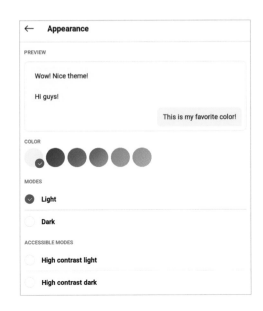

10 Tap on the **Calling Settings** setting to select options for making and receiving voice calls, including displaying caller ID and using subtitles for a call

11 Tap on the **Messaging** setting to select options for text messaging (including using read receipts, so that you know when someone has read a message) and automatically downloading photos in a message

12 Tap on the **Notifications** setting to select options for how you are notified regarding chat messages

If you are going to be downloading photos from a text message in Skype it is better to do this over Wi-Fi, rather than a smartphone data connection. This is because it will use up a certain amount of your data allowance (if it is not unlimited) and you may have to pay additional charges if you go over your data allowance.

...cont'd

Hot tip

Tap on the **Sync your contacts** option in Step 13 to synchronize your contacts in the address book on the device with which you are accessing Skype, with your Skype account.

13 Tap on the **Contacts** setting to select options for synchronizing the contacts on your device, blocking contacts, and privacy issues

14 Tap on the **Help and feedback** setting to select help options for finding out more information about Skype. Tap on the **Get help** option

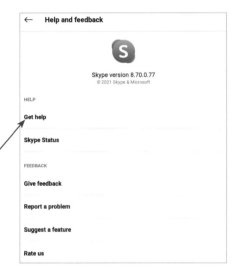

15 The **Get help** option provides an online help page, where you can enter specific keywords in the **Search** box, or tap on the icons below to view help information about these topics

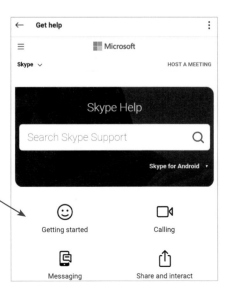

Editing Your Status

The account icon at the top of the Skype window is used to display the user's status, which can be customized to show different options. This can let other users know if you are available for a call or are away. To change your own status:

1 Tap on your account

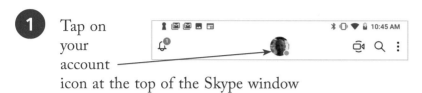

icon at the top of the Skype window

2 Your current status is displayed

Your status is what other Skype users will see when they try to contact you.

3 Tap on your current status to view the available options

4 Tap on the required status to apply it

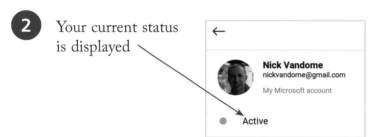

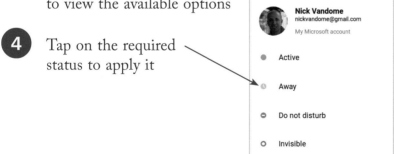

Tap on someone's account icon, or move the cursor over it, to view a description of their status.

5 The current status is applied to your account icon at the top of the Skype window

107

Adding Contacts

There are two main ways to add contacts to Skype so that you can then communicate with them: adding contacts manually, or connecting with people who are already using Skype and inviting them to be a contact.

Adding contacts manually
To add contacts to Skype manually:

Existing contacts in your address book can also be synchronized with Skype, as shown in the Hot tip on page 106.

For the desktop and laptop versions of Skype, the **+ Contact** button is at the top of the Contacts window.

1 Open Skype and tap on the **Contacts** button on the bottom toolbar

2 Tap on the **My contacts** tab at the top of the window

3 Tap on this button in the bottom right-hand corner of the window

4 Tap on the **Invite to Skype** button

Other ways to add people

S **Invite to Skype**
Invite friends to Skype via SMS, email and more.

☎ **Add a phone number**
Save a number to your Skype contacts.

5 Select an option for how you would like to invite someone to view your Skype profile

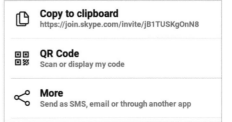

⊕ **Copy to clipboard**
https://join.skype.com/invite/jB1TUSKgOnN8

▦ **QR Code**
Scan or display my code

⤳ **More**
Send as SMS, email or through another app

The **QR Code** option in Step 5 displays a unique code image, which someone else can scan, to be added to your contacts.

6 For the **Copy to clipboard** option in Step 5 on the previous page, create the required item (e.g. an email) and paste the link into an email. The recipient will be able to use this to access your Skype account

For the desktop and laptop versions of Skype, the **Invite to Skype** options in Step 5 on the previous page are only **Copy to clipboard** and **Email**.

7 For the **More** option in Step 5 on the previous page, select one of the options for inviting someone to access your Skype account

The icons that appear in Step 7 will vary depending on the apps on your device.

8 For an item such as an email, the link will be pre-inserted into the email

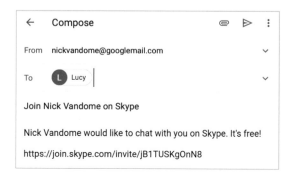

...cont'd

Adding Skype contacts

To add other existing Skype users as contacts:

Beware

If the search item in Step 2 is not recognized or there are no matches, there will be no suggested contacts displayed.

1 Access the **Contacts** section and tap on the new contact button, as shown in Step 3 on page 108

2 In the **Search** box at the top of the window, enter the name of a person, an email address, a phone number, or their Skype username, if known

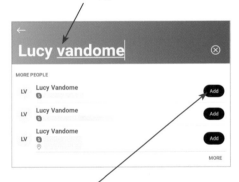

3 Tap on the **Add** button to add someone as a Skype contact. Once this has been done, the button changes to **Added**

4 The person is added to the **Contacts** section

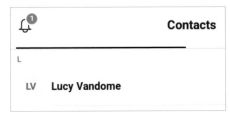

Making a Test Call

Before you start making video and audio calls to your family and friends who you have added to Skype, you can make a test call to make sure everything is working. To do this:

1 Access the **Contacts** section, scroll down to the **S** section, and tap on the **Skype Test Call** button

2 Tap on the phone icon to connect to the Skype Test Call window

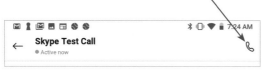

3 The Skype Text Call window will ask you to record a short message, when the tone sounds and the main icon pulses blue. The message will be played back to you, and if you can hear it then your Skype app is working correctly

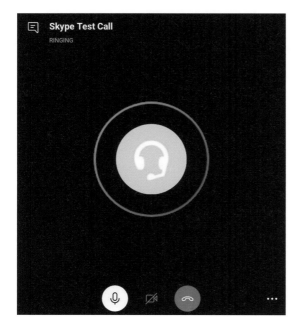

Make sure that your microphone is turned **On** for the test call.

The test call is just for audio, not for the video aspect of Skype: the video button is automatically disabled.

Starting a Video Call

Video calls in Skype can be instigated from all of the main sections of the app.

Video calls from the Contacts section
To start a video call from the Contacts section:

1 Access the **Contacts** section, and tap on the person with whom you want to have a video call

2 Tap on the video icon to make a video call

3 Once the recipient has accepted the call, their video feed appears in the main window and yours appears as a thumbnail at the top of the window

Tap on the phone icon in Step 2 to make a voice-only call.

Video calls from the Calls section
To start a video call from the Calls section:

1 Tap on the **Calls** button

2 All of the **RECENT CALLS** are listed

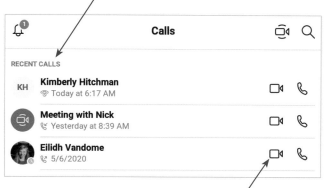

If you tap on a person's name in Step 2, the options for a video or audio call are displayed in a panel at the bottom of the screen.

3 Tap on these buttons next to someone in the **RECENT CALLS** section to call them again, with a video or an audio call

4 Your contacts are listed below the **RECENT CALLS** section, and these can be called in the same way as in Step 2

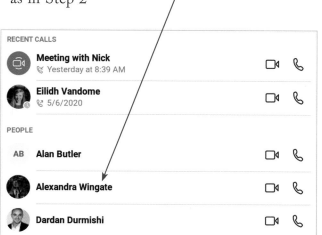

...cont'd

Video calls from the Chats section

To start a video call from the Chats section:

1 Tap on the **Chats** button

2 The Chats section lists all of the video and audio calls that been made and received

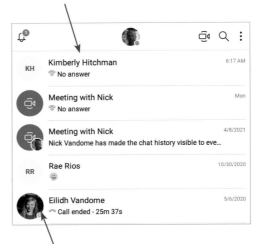

The Chats section contains details of text chats you have had with family and friends, and also requests from people to add you to their Skype account.

114

3 Tap on a person to view details of the related item – e.g. a call or a chat

4 Use these buttons to make a video or an audio call

5 The Chats section also contains a log of calls that have been made with the selected person, and a text box at the bottom of the window, for text chats

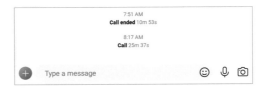

Creating a Group Call

Another option for starting a video call is to create a group meeting from the main toolbar. This can be used to create a recurring meeting (chat) with a group of people. To do this:

1 In either the **Chats** or the **Calls** section, tap on the video icon on the top toolbar

2 A new meeting window appears, for creating the elements of the meeting

115

Hot tip

Group calls with other Skype users are unlimited – e.g. you can stay in the call for as long as you like.

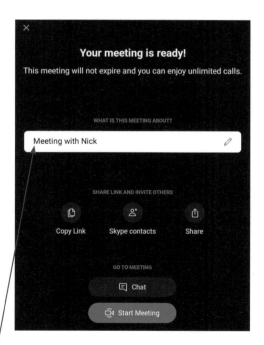

3 Tap on the text box and change the title of the meeting, as required

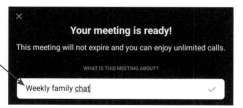

...cont'd

4 Tap on the **Skype contacts** button to add people from your existing Skype contacts

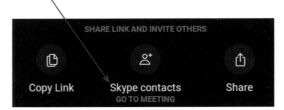

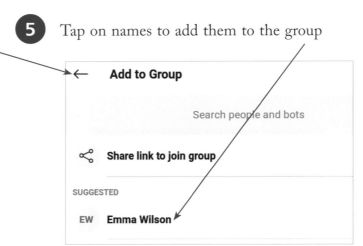

Don't forget

Tap on the arrow next to the **Add to Group** heading to return to the previous window, without adding any new users.

5 Tap on names to add them to the group

6 The names that have been added appear at the top of the **Add to Group** window

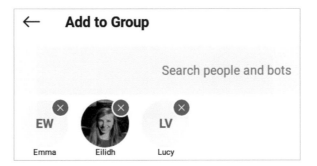

7 Tap on the **Done** button once all of the required people have been added

...cont'd

8 To view the details of a group call, tap on the **Chats** button

Chats

9 Tap on the name of a group that has been used in a previous call

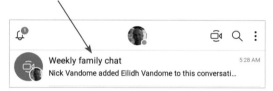

10 Tap on the **Start call** button to begin another chat with the group. Tap on the **Add people** button to add more participants to the group chat

11 Tap on the group heading in the previous step to view details of the group. Tap on items, such as: **Start call**, to start a new call to the group; **Send message**, to send a text message; or **Schedule a call**, to set a time for a new call to the group

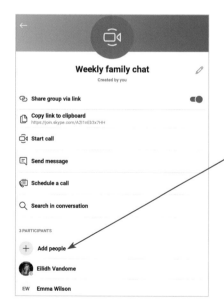

Hot tip

Tap on the **Add people** button in Step 11 to add more people to the group, in the same way as when adding them initially.

During a Video Call

Once a one-to-one video chat has been connected, by whatever method, there is a range of options that can be applied during it:

Depending on how another participant on a call is using their video feed (e.g. in landscape or portrait mode on their device), this will determine how the video feed appears on your own screen.

1 By default, your own video feed is a thumbnail at the top of the screen and the other participant's video feed appears in the main screen area

2 Drag your own video feed to the bottom of the window to resize both feeds to the same size

...cont'd

3 Drag your own video feed back to the top of the window to reinstate it as a thumbnail

Beware

You cannot move someone else's video feed by dragging it. It will be resized if you move your own one but, for instance, you cannot drag it into a different position on the screen.

4 Tap on the screen to access the video controls, at the top and bottom of the screen

Don't forget

The duration of the call is shown in the top left-hand corner of the screen.

...cont'd

Beware

If you turn your speaker Off, you will not be able to hear the other participant(s) in a video or an audio call.

5 Tap on this icon on the top toolbar to send a text message to the participant during the video call (see the next page)

6 Tap on this button to turn your speaker **Off** or **On**

7 Use the four buttons on the bottom toolbar to, from left to right: add someone as a favorite; mute or unmute your microphone; turn your video feed **On** or **Off**; and end a call

Hot tip

Use the buttons at the bottom of the window in Step 8 to: add more people to the call; record a video chat; take a photo of the entire screen; and share your screen with the other participant(s) – i.e. open another app, such as the Photos app, to show people what is displayed in that app.

8 Tap on this menu button to access additional options for the call, including: blurring the background of your video feed; allowing incoming video; and using subtitles on a video call

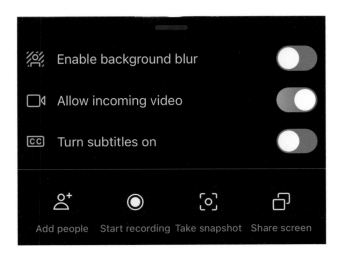

Sending a text message during a call

It is possible to send someone a text message while you are in a video chat with them. To do this:

1 Tap on the button in Step 5 on the previous page to access the **Chats** window for the person in the call

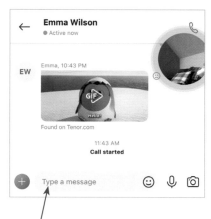

2 Enter a text message and tap on the **Send** button. If you receive a text message during a video chat it will appear briefly on screen, with a blue background

The text message in Step 2 will disappear after a few seconds, but it will still be available in the **Chats** section, under the person's name.

Adding Text Chat

In addition to using Skype for video chatting, it is also extremely effective for text chats. This can be done as a stand-alone conversation, and also during a video chat. To get started with text chatting in Skype:

1 Tap on the **Chats** button

2 Tap on this button to start a new chat conversation

3 Tap on one of the options for connecting with people, including starting a video call (**Meet Now**) or an audio call (**New Call**), or for a variety of options for text chatting, or tap on a contact's name to chat directly to them

122

The text-chat options in Step 3 include: **New Group Chat**, which can be used to set up a regular group with which you want to chat; **New Moderated Chat**, which is another group-chat option, but one in which members can be authorized to monitor what is said in the chat group; and **New Private Conversation**, which is a conversation that is designed to be as secure as possible, by using end-to-end encryption.

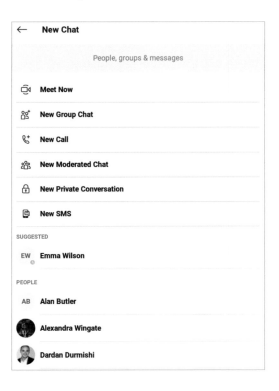

4 If you select an existing contact, the **Chats** window will open, with any previous conversations and messages displayed

5 Tap in the **Type a message** text box at the bottom of the window

6 Enter a message and tap on the **Send** button

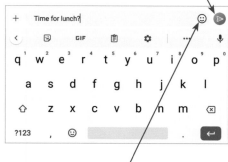

7 Tap on the emoji icon in the text box to access options for adding **Emoticons** (animated emojis), animated **GIFs**, **Stickers** and **Mojis**

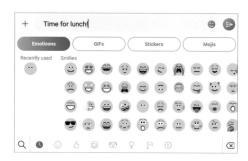

The **Chats** interface for the Windows and Mac versions of Skype is slightly different from the mobile version. The **Chats** text box has an emoji icon to the left of the text box. To the right are options for, from left to right: adding files; adding contacts to a chat; starting a video call; and accessing the menu of additional options.

Animated GIFs and Mojis are similar: they are both animated images, usually used for humorous effect.

8 To add a photo to a text message, tap on this button next to the text box

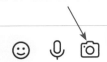

9 Tap on the **ALLOW** button to enable Skype to access your photos and other files to add to a text chat

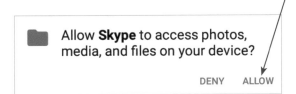

Hot tip

The options in Step 10 can also be used to take photos with your device's camera to add to a text message, and add videos or documents.

124

10 Tap on a photo to select it, and tap on this button to send it in a text chat

Hot tip

Several photos can be added in Step 10, by tapping on all of the ones that are required.

11 The selected photo(s) is added to the text chat

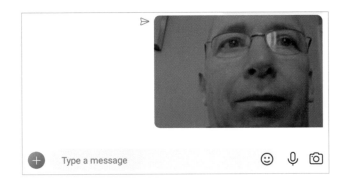

12 The toolbar below the text box contains options for adding animated icons and images, items that have been copied to the Clipboard, and for accessing the settings for the keyboard

13 Tap on this menu button to access more options for working with text messages, including: changing the **Theme** for the Chats window; using a **One-handed** keyboard; **Text Editing**; **Translate** text; and using a **Floating** keyboard that can be moved around the screen

14 Tap on the **+** icon to the left of the text box to access options for adding **Content and tools** to a text chat. This includes content such as photos, videos and files, and also options for sharing contacts and your location via a text message

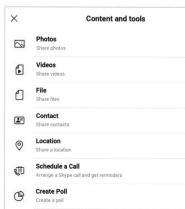

Tap on this button to the left of the toolbar in Step 12, to hide the toolbar:

The options in Steps 12 and 13 are only available in the Android version of Skype.

In a private conversation (see the Don't forget tip on page 122) the items in Step 14 are limited to the **Photos**, **Videos** and **File** options.

Ending a Skype Call

To end a Skype call of which you are the host – i.e. you have started:

1 During the call, tap on the screen and tap on this button at the bottom of the screen

2 For a one-to-one video call, this ends the call for both people. For a group video call, the other people in the call will still be connected, and only your connection will be ended

Declining a call

If a call is declined, there is an option for sending a text message instead:

1 Someone can decline a call by tapping on this button, or by not answering the call

2 If a call is declined or not answered, you are notified that the person is unavailable. A text message is pre-inserted to send

3 Tap on the **Send** button to send the text message. This appears on the relevant **Chats** page

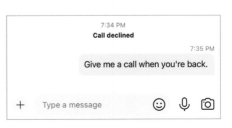

Hot tip

If a call is declined, you can send a customized text message: double-tap on the pre-inserted text and write your own message.

7 FaceTime

FaceTime is Apple's own video-chatting app that can be used on its digital devices, from the iPad to the iMac. This chapter covers all of the issues and features to use FaceTime for successful video chats.

FaceTime is a good option for video chatting between Apple users. However, because it cannot be used on other devices, if someone has a non-Apple device then using another video-chatting app is a better option.

Even if you remove the FaceTime app, this functionality will still remain on the device. For instance, you will be able to receive FaceTime calls, and also make them from the Contacts app, by tapping on the **FaceTime** button next to a contact (provided they have an Apple ID).

About FaceTime

FaceTime is Apple's video-chatting app that is used on all Apple devices: iPhones, iPads, iMacs, MacBooks, Mac minis, and Mac Pros. It is only available on these devices and cannot be used on other platforms.

FaceTime comes pre-installed on Apple devices. It is ready to use but requires an Apple ID, which can be created when you first start using an Apple device or when you first open an app that requires this. An Apple ID can also be created on the Apple website at **https://appleid.com/account**, using a username or email address, and a password.

Although FaceTime comes pre-installed on Apple devices, it is possible to remove the app – for instance, if you want to use other video-chatting apps. However, if FaceTime is removed, it can be reinstated from the Apple App Store. To do this:

1 By default, the FaceTime app is on the Home screen of an iPad and iPhone, and within the Launcher of an Apple desktop or laptop

2 To remove the FaceTime app, press on the app until a small cross appears in the top left-hand corner. Tap on the cross

3 Tap on the **Delete** button to confirm that you want to remove the FaceTime app

Delete "FaceTime"?

You will still be able to make and receive FaceTime calls.

Cancel | Delete

Downloading FaceTime

If FaceTime has been removed from an Apple device, it can be downloaded again from the App Store. To do this:

1 Tap on the **App Store** app

2 Tap on the **Search** button

3 Enter "facetime" into the Search box and tap on one of the results

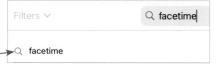

4 Details about the FaceTime app are displayed. Tap on the cloud icon to download it (this icon is used since the app has already been downloaded and is available to download again)

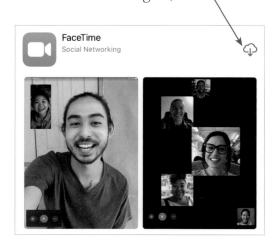

5 Tap on the **OPEN** button to open the FaceTime app, or open it from the Home screen

The examples in this chapter are from FaceTime on an iPad, using the iPadOS operating system, except for pages 144-145, which show how to add special effects with FaceTime, using an iPhone.

All other major video-chatting apps can also be used on Apple devices.

If you reinstate the FaceTime app after it has been removed, it will not be replaced in its original position on the Home screen; it will instead appear at the end of all of the available Home screens.

FaceTime Settings

Unlike with some other video-chatting apps, the settings for FaceTime are accessed from a separate Settings app, rather than from within the FaceTime app itself. To view these:

1 Tap on the **Settings** app

2 Tap on the **FaceTime** tab

3 The FaceTime settings are displayed

4 Drag the FaceTime button **On** (displays green) to ensure that FaceTime is active on the related device

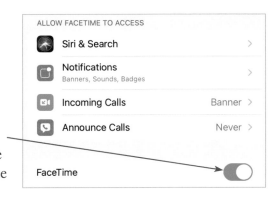

ALLOW FACETIME TO ACCESS

Siri & Search >

Notifications
Banners, Sounds, Badges >

Incoming Calls Banner >

Announce Calls Never >

FaceTime

5 Tap on the **Notifications** button in Step 4 to determine how you will be notified when you receive a FaceTime call

< FaceTime Notifications

Allow Notifications

ALERTS

Lock Screen   Notification Center   Banners

Banner Style   Temporary >

Sounds   Opening >

Badges

Don't forget

Mac computers using macOS do not have specific settings for FaceTime. However, **System Preferences > Notifications** can be used to specify how notifications for FaceTime calls are handled, in a similar way to Step 5.

6 Tap on the **Incoming Calls** button in Step 4 to specify what happens when you receive a FaceTime call while the device you are using is locked

7 Tap on the **Announce Calls** button in Step 4 to specify circumstances for an audible notification when you receive a FaceTime call

8 Swipe down the page in Step 4 and tap on the **Respond with Text** button

9 Enter the text you want to display if you cannot answer a FaceTime call

If your device is locked – i.e. you are not using it – an incoming FaceTime call will take up the entire Lock screen.

See pages 136-137 for more details about accepting and answering FaceTime calls.

The grayed-out options in Step 9 are just suggestions, but will not appear unless you enter this text.

Accessing Contacts

Video calls can be made with FaceTime to existing contacts that you have on your Apple device, or new ones can be added. To do this:

The list of calls in Step 2 includes video and voice calls that have been made on all linked Apple devices – i.e. those that share the same Apple ID.

1 Tap on the **FaceTime** app to open it

2 Calls to previous contacts are listed in the left-hand panel. Tap on one to start a new FaceTime call

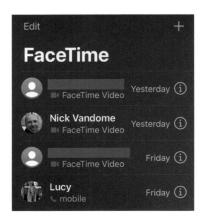

3 Tap on this button at the top of the left-hand panel to start a new call

4 Tap in the **To:** box and start typing a name. As you type, options appear below the **To:** box. Tap on an option to select it

5 Tap on the **Audio** or **Video** button at the bottom of the left-hand panel to start one of these calls

Starting FaceTime calls with contacts

To start a FaceTime call with someone in your address book
(Contacts app):

1 Tap on the
+ icon in
the Contacts
window (as
shown in
Step 4 on the previous page)

2 The contacts
from the
Contacts app
are displayed.
Tap on one to
select someone

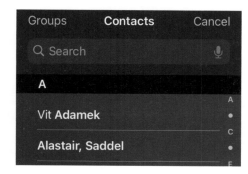

3 Once someone
has been
selected, tap
on the **video**
button

4 If you have
more than one
video-chatting
app on your
device, tap on
the **FaceTime**
option

Hot tip

If you do not have any
other video-chatting
apps on your device,
the button in Step 3
will say **FaceTime**
rather than **Video**.

Starting a FaceTime Call

To start a video call on FaceTime, to another FaceTime user on an Apple device:

1 Access a contact, as shown on pages 132-133, and select one

2 The call is made to the other person

Your own thumbnail video feed can only be positioned in one of the four corners on the screen. It cannot be positioned over other video feeds.

3 When the other person accepts the call, their video feed is in the main window, and yours is a thumbnail in a corner of the screen

4 Press on your video thumbnail and drag it to a different position on screen, as required

5 The FaceTime video chat controls panel is located in the bottom left-hand corner. Tap on the screen to hide, or show, the controls panel

6 Swipe on the bar at the top of the controls panel to expand it and view additional options for the call

Don't forget

The position of your own video thumbnail will be retained if you close FaceTime and then open it up again – i.e. the thumbnail will remain in the same place as before the app was closed.

Hot tip

Swipe down on the bar at the top of the controls panel in Step 6, to return it to its original size.

Receiving a FaceTime Call

When you receive a FaceTime call, there are slightly different options, depending on whether your device is locked or not:

If you tap on the red **Decline** button in Steps 1 or 2, the call will be declined immediately, with no other options.

1 If your device is unlocked, the notification for the FaceTime call appears as a banner at the top of the screen. Tap on this button

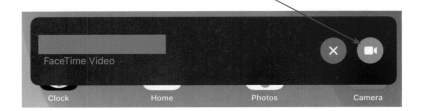

2 The FaceTime call is expanded to fill the whole screen, with the caller's name or number at the top of the screen

3 Tap on the **Accept** button in Step 2 on the previous page to start the FaceTime call, or tap on the **Decline** button to not accept it

4 Tap on the **Message** button in Step 3 to send a text message, rather than accepting or declining the call

Custom...

Can I call you later?

I'm on my way.

Sorry, I can't talk right now.

Hot tip

If you have created a **Respond with Text** option in the FaceTime settings, as shown on page 131, this will be available on the list in Step 4. Tap on the **Custom** option to create a specific text message to the caller.

5 Tap on the **Remind Me** button in Step 3 to decline the call but create a reminder in 1 hour about the call

Remind Me Later

In 1 hour

6 If your device is locked, the FaceTime call screen uses the **slide to answer** button option for accepting the call (there is no option for declining the call)

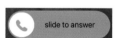

slide to answer

Managing a FaceTime Call

During a FaceTime video chat, there are various options for how you can manage the call:

1 Use these buttons in the video controls panel to, from left to right: turn your own video feed on or off; mute, or unmute, your microphone; flip your camera, between the front-facing and back-facing ones; and end the call

2 If the camera is turned off and the microphone muted, both of these icons appear with a white background

3 If the camera is flipped, it changes from the front-facing camera (which is used to display yourself) to the back-facing one, which displays whatever the camera is pointed at

Flipping the camera can be a good option if you want to show someone an item such as a notable landmark or an impressive sunset.

4 Press the **Home** button to minimize the FaceTime window to the top right-hand corner

Hot tip

The video chat can be continued if the FaceTime app has been minimized into a thumbnail, as in Step 4.

5 Tap on the cross in the top left-hand corner of a thumbnail to hide it, and pause your video feed. Tap on this button to resume the chat

6 If a video chat has been paused, this is displayed in the participant's video-feed window

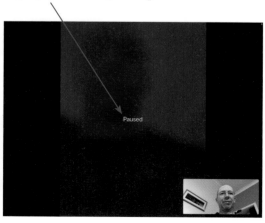

Making a Group Call

FaceTime video chats are not limited to one-to-one options: up to 32 people can be included in a video chat. To add more people to an existing video chat with two people:

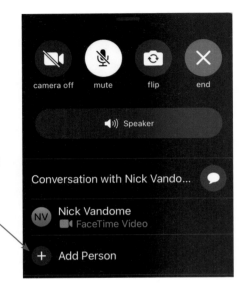

1 During a video chat, expand the video controls panel, as shown on page 135, and tap on the **+ Add Person** option

2 Enter someone's name here and tap on one of the results

Hot tip

A full name does not need to be entered in Step 2: as you start typing, options appear based on the letters that have been entered; the more letters that are included, the more defined the results.

3 Tap on the **Add Person to FaceTime** button to add the selected person to a call

4 Tap on the **+** button at the right-hand side of the **Add Person** window in Step 2 to access the Contacts app for adding people from here

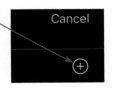

...cont'd

5 Tap on a contact to add them to a chat, as shown on page 133

6 By default, during a group call, all of the video-feed windows float across the screen. While the calls are connecting, the video feeds are displayed as icons

7 Tap on a video feed to make it the active one and tap on this icon to expand it on screen

Beware

If there are a lot of people in a FaceTime call, the floating windows can be distracting.

...cont'd

8 The selected video feed is maximized on the screen, with the other video feeds appearing behind it

Hot tip

Groups that have been created in the Contacts app can also be used to initiate a group FaceTime chat.

9 Tap on this button to minimize a window that has been maximized

10 Once a group call has been made, the group is listed in the calls list. Tap on the group to call all of the people in the group again

Edit +

FaceTime

Nick Vandome or 2 oth...
Tap to Join

Taking a Photo

During a FaceTime call, it is possible to take a photo of someone else's video feed. To do this:

1 Start a call and, at the required point, tap on this button to take a photo of the other person's video feed

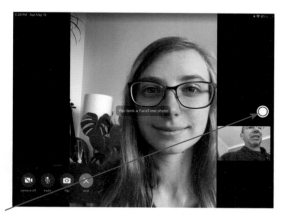

2 The **You took a FaceTime photo** notification appears once you have captured a photo

3 Tap on the **Photos** app to view the photo that has been taken in FaceTime

4 The photo is displayed within the **Photos** app interface

It is not possible to record a video of what is taking place in a FaceTime chat; you can only take a photo.

If you take a photo of someone during a video chat, they will receive a notification about this on their own screen.

Adding Special Effects

Some iPhone models have a FaceTime option on the video controls panel for adding special effects to your own video window. To use this:

1 During a video chat, tap on this button on the video controls panel

2 The special-effect options are displayed in a toolbar above the video controls

Don't forget

Once a special effect has been applied, this is what everyone else in the FaceTime call will see.

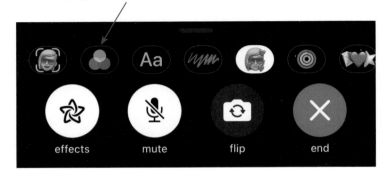

3 Tap on this button to add a Memoji effect to your video feed. This is an animated effect that copies your facial expressions

4 Tap on this button in Step 2 on the previous page to add colored filter effects

5 Tap on this button in  Step 2 on the previous page to add text to your video feed

Special effects are best used sparingly, as they can become annoying if there are too many of them.

6 Tap on this button in Step 2 on the previous page to add Memoji stickers. Tap on the stickers to add them to your video feed

Text Chat During a Call

Text messages can be sent to individuals during a FaceTime video chat. To do this:

1 During a video chat, swipe up on the video controls panel to expand it

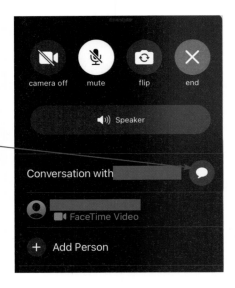

2 Tap on the **Conversation with** icon

If you have had any previous text chats with the person in the video call, these will be displayed in the **Messages** window in Step 3.

3 The Messages app opens in the message window for the person in the FaceTime video chat. The FaceTime window is minimized in the top right-hand corner of the screen

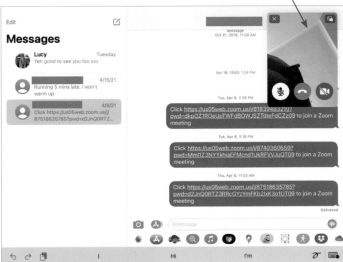

4 Enter a text message in the text box at the bottom of the screen and tap on the **Send** button

5 The message is added to the text conversation with the person in the FaceTime call. As more text messages are added, they are displayed within the Messages app. The FaceTime call can continue while messages are added

6 Tap on this button to the left of the text box to access options for adding more content to a text conversation

7 Tap on this button to add photos to a conversation. Tap on a photo(s) to select it and tap on the **Add** button to add it to a text conversation

147

Ending and Calling Back

The process for ending a FaceTime call is the same whether you have made the call or received it. If a call is declined, it is possible to call the person back or leave a text message for them.

Ending a call

To end a FaceTime call:

1 During the call, tap on the screen to access the video controls panel and tap on the **end** button

Calling back

To call someone back or leave a message, if a call has not been connected or has been declined:

The **Call Back** and **Leave a Message** options are available whether someone actively declines a call or there is just no answer.

1 If a call is not connected, for whatever reason, tap on the **Call Back** button to call the person again

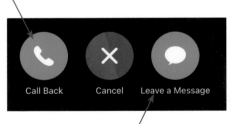

2 Tap on the **Leave a Message** button to send the person a text message

3 The **Messages** app opens automatically. Type the required message and tap on the **Send** button

8 Messenger

Messenger is a popular Facebook app for text and video chatting. This chapter covers everything you need to know to get up and running with it.

The terminology used in this chapter refers to tapping on a touchscreen device. For versions of the app on a desktop or laptop computer, the equivalent action will be clicking. The examples in this chapter are from Messenger on an Android tablet.

A Facebook account is needed to use Messenger (although you can deactivate your Facebook account once created, and continue to use Messenger), and these details are used to log in to Messenger. If you already have a Facebook account you can use your login details to use Messenger, or create an account when you first start to use it.

Obtaining Messenger

Messenger is owned by Facebook and is now a stand-alone app, after formerly just being incorporated into Facebook.

Messenger for iPhones and iPads

To download Messenger for iPhones and iPads, the process is the same for both:

1 Tap on the **App Store** app

2 Access the Messenger app and tap on the **GET** button to download it

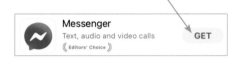

3 Tap on the **Messenger** app to open it

Messenger for Android smartphones and tablets

To download Messenger for smartphones and tablets using the Android operating system:

1 Tap on the **Play Store** app

2 Access the Messenger app and tap on the **Install** button to download it

	Messenger – Text and Video Chat for Free
	Facebook In-app purchases
	4.2★ 77M reviews 1B+ Downloads Editors' Choice Parental guidance ⓘ
	Install

3 Tap on the **Messenger** app to open it

...cont'd

Messenger for Windows 10 devices

For users of Windows 10 devices, Messenger can be downloaded from the Microsoft Store. To do this:

1. Click on the **Microsoft Store** button

2. Access the Messenger app and click on the **Get** button to download it

Messenger
Facebook Inc • Social
↪ Share
Made for big screens and close connections. video chat built specifically for desktop.
More

EVERYONE
Users Interact
ESRB

Free
Get

3. Open the Start menu and click on the **Messenger** app to open it

Messenger for macOS devices

To download Messenger on Apple computers using macOS:

1. Click on the **App Store** app

2. Access the Messenger app and click on the **GET** button to download it

Messenger
Text, Voice, & Video chat → GET

3. Click on the **INSTALL** button to install the Messenger app

INSTALL

4. Open the **Launcher** and click on the **Messenger** app to open it

Hot tip

To access the **Launcher** (displaying all of the apps on the device) on a Mac desktop or laptop, click on this button on the Dock at the bottom of the screen:

Messenger Settings

There is a range of settings for Messenger so that you can customize it to your own preferences. To do this:

Hot tip

Drag the **Dark Mode** button in Step 3 **On** to invert the screen background and view white text on a black background.

Hot tip

Tap on the **Active Status** button in the **Profile** section and drag the **Show when you're active** button **On**, so that your Messenger contact will be able to see when you are online and using Messenger. If you do not want people to see this, drag the button **Off**.

1 Tap on the **Messenger** app to open it

2 Tap on your account icon, in the top left-hand corner of the screen

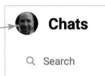

3 The main settings categories are displayed, which cover **Profile**, **Preferences** and **Account**. Tap on an item to view the sub-categories within it

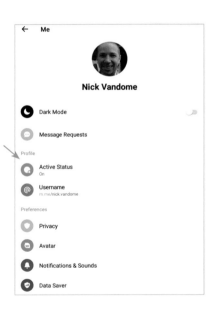

4 Swipe up the screen to view the full range of **Preferences** settings. These include options for **Privacy**; **Notifications & Sounds**; **Phone Contacts**; and **Photos & Media**

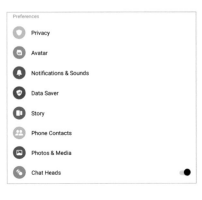

5 Tap on the **Privacy** option in Step 4 on the previous page to access settings for keeping your Messenger interactions as secure as possible

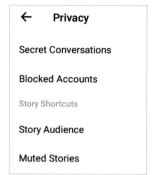

6 Swipe up the screen in Step 3 on the previous page to view the full range of **Account** settings

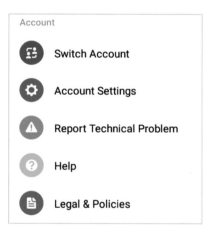

The **Help** option in Step 6 takes you to the online Facebook Help Center, which has a wealth of information about using Messenger and also Facebook.

7 Tap on the **Account Settings** option in Step 6 to access settings for privacy options, how advertisements are handled and how your personal data is used

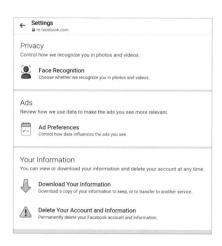

If the Messenger interface looks significantly different on your device than in the illustrations here, further help is available on our website. Visit **www.ineasysteps. com/products-page/ video-chatting-for- seniors-in-easy- steps** and click on Resources.

Accessing Contacts

Messenger can be used with your existing Facebook contacts, and also contacts from your address book on the device on which you are using Messenger. To access contacts:

1 Tap on the **Messenger** app to open it

2 Tap on the **People** button on the bottom toolbar

The button on the left-hand side in Step 3 is the **All People** button.

3 Tap on this button in the top right-hand corner to view all of your Facebook contacts

Facebook contacts are automatically available in Messenger; they do not need to be added.

4 All of your Facebook contacts are listed. Tap on one to connect to them for a video chat

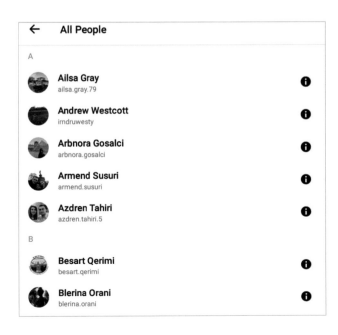

5 Tap on this button in the top right-hand corner to add contacts from your device's address book

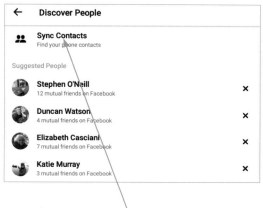

6 Tap on the **Sync Contacts** button

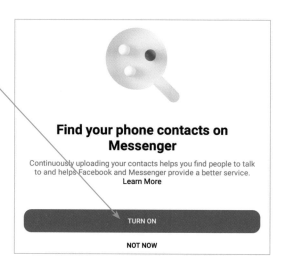

7 Tap on the **TURN ON** button on the bottom toolbar, to add your device's contact to Messenger

Find your phone contacts on Messenger

Continuously uploading your contacts helps you find people to talk to and helps Facebook and Messenger provide a better service. Learn More

TURN ON

NOT NOW

8 Tap on the **ALLOW** button to enable Messenger to access your device's contacts

Allow **Messenger** to access your contacts?

DENY   ALLOW

Hot tip

The screen in Step 5 also lists suggested people, based on your existing Facebook friends. These are people with whom you have mutual friends – i.e. you both have at least one Facebook friend in common. Tap on one of the suggested people to add them as a friend.

Setting Up the Camera

In order for Messenger to be able to conduct video chats, the device's camera has to be set up for use with the Messenger app. To do this:

1 Open Messenger and tap on the **Chats** button on the bottom toolbar

2 Your own account icon is displayed at the top of the window. Tap on the camera icon to set up the camera for use with Messenger

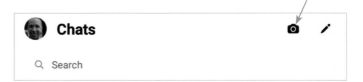

Beware

If you do not allow Messenger access to your camera, you will not be able to take part in any video calls.

3 Tap on the **TURN ON** button

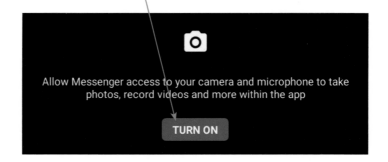

4 Tap on the **ALLOW** button to give Messenger access to your device's camera

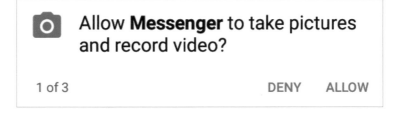

Tap on the **ALLOW** button to allow Messenger to record audio calls

🎤	Allow **Messenger** to record audio?
2 of 3	DENY ALLOW

 Tap on the **ALLOW** button to give Messenger access to your device's photos and other items

📁	Allow **Messenger** to access photos, media, and files on your device?
3 of 3	DENY ALLOW

7 The device's camera is opened, indicating that Messenger has been given access to it

The camera view in Step 7 is the same one that will be used for all video calls.

Starting a Video Call

Video calls, and other communication such as an audio call or a text chat, can be made to any of your contacts in the Messenger app. This then forms part of a record of communication with the selected person. To start a video call with one of your Messenger contacts:

1 Tap on the **People** button on the bottom toolbar

2 Access the **All People** section, as shown on page 154, and tap on one of your contacts

3 The buttons on the top toolbar contain options for, from left to right: starting an audio call; starting a video call; or viewing details about the selected person

Hot tip

A video call can be started by tapping on the video icon in Step 3; the details in Step 4 can be used to view more information about someone, and then contact them in various ways.

4 Tap on the **i** icon to access more details about them, such as photos and videos that they have posted on Facebook, and other conversations in which their names have appeared, other than ones with you

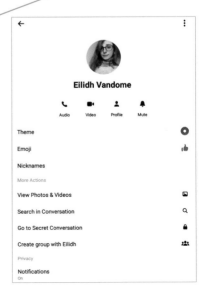

5 Tap on the **Video** button in Step 4 on the previous page to start a video call with the selected person

Video

6 Once the selected person has accepted the call, it is connected. Their video feed appears in the main window, and your video feed is a thumbnail in the corner of the screen

Hot tip

Once you have had a video chat with someone, details of this will be displayed in the **Chats** section, under the person's name. Tap on the person's name here to access details of your communications with them, and add to it, as required.

159

7 Tap on the screen to access the video control buttons

Managing Video Calls

When a video call has been connected to another Messenger user, there are several options for managing it:

Hot tip

If a photo is taken during a video call, it will be stored in the device's default Photos app.

Don't forget

Tap on the **ADD PEOPLE** button in Step 3 to add more people to the current video call.

1 With the video controls showing, tap on this button to take a photo of the other person in the video call

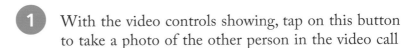

2 Use the video control buttons to, from left to right: show, or hide, your own video feed; switch the camera between front-facing and back-facing; mute, or unmute, the microphone; and end the call

3 Swipe up on the small bar at the top of the video controls to access more options, including adding more people to the call and sharing your screen

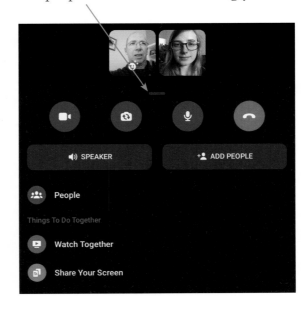

Adding effects

Special effects can also be added in a Messenger call:

1 Tap on this icon in your own video-feed window

2 The effects bar is displayed at the bottom of the window

3 Swipe along the bar to access the different categories

Backgrounds and effects can be changed or removed while a call is taking place.

4 Tap on an item in one of the categories to apply it to your video feed

Creating Groups

Creating groups in Messenger is a good way to contact several people at the same time, particularly if you communicate with them regularly. To create a group:

1 Tap on this button on the top toolbar

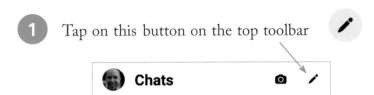

2 Tap on the **Create a New Group** button

Don't forget

If someone is added to a group, they do not have to accept an invitation; they are added to the group automatically. However, they can leave the group, if required.

3 On the **Add Participants** page, tap on a name to add them to the group. Tap on the **NEXT** button

4 Enter a name for the group in the **Name your new chat** text box and tap on the **CREATE** button

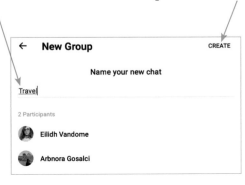

5 The conversation page for the group is displayed. Use the buttons on the top toolbar to start an audio or video call to the group, in the same way as for an individual

6 To find a group, tap on the button in Step 1 on the previous page and start entering the group name. Tap on one of the results

7 Tap on the **i** icon on the top toolbar in Step 5 to view details about the group, and options for managing it

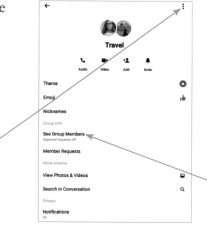

Hot tip

To remove someone from a group you have to be the group administrator, usually the person who has set up the group. If this is the case, tap on the **See Group Members** option in Step 7, tap on a member in the group and tap on the **Remove From Group** button.

8 Tap on the menu button in Step 7 to access more options for working with the group, including one for leaving the group

Text Chatting

As well as having audio and video chats in Messenger, it is also possible to have text chats with individuals or groups. To do this:

1 Select an individual, or a group that you have created or been added to. The conversation page opens, showing any communication you have had with the selected person/group

2 The text box and its options are located at the bottom of the conversation window

Don't forget

The keyboard display changes depending on the characters being entered in a message. Tap on this button to access a different keyboard:

?123

3 Tap in the text box to activate the keyboard, ready for typing a message

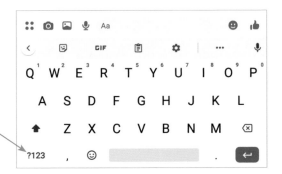

4 Enter a message in the text box, as required

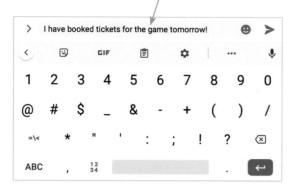

Tap on this button to the left of the text box to show the buttons to the left of the text box, as in Step 2 on the previous page:

> I have

5 Tap on the emoji icon in the text box to access options for adding **Stickers**, **GIFs** and **Emojis**

Tap on the emoji icon again to hide the emoji window and return to the standard keyboard option.

6 Tap on an emoji to add it to the message

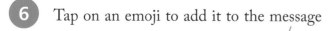

7 To take a photo to add to a text message, tap on the camera button to the left of the text box.
Use the camera to take a photo or record a video that will be added to the text conversation

Tap on this button to send a message:

...cont'd

8 The toolbar below the text box contains options for: adding animated icons and images; items that have been copied to the Clipboard; and accessing the settings for the keyboard

Tap on this button to show or hide the toolbar in Step 8:

9 Tap on one of the options on the toolbar in Step 8 to view it. Tap on the **ABC** button to return to the standard keyboard

10 Tap on this menu button in Step 8 to access more options for working with text messages, including: changing the **Theme** for the Chats window; using a **One-handed** keyboard; **Text Editing**; **Translate** text; and using a **Floating** keyboard that can be moved around the screen

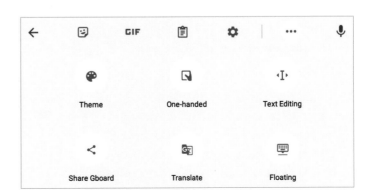

9 Facebook Portal

Facebook Portal is a touchscreen device that is designed specifically for video chatting with the Messenger and WhatsApp apps. This chapter shows you how.

About Facebook Portal

Facebook Portal is a touchscreen display that is designed specifically for video calls. However, it can also perform other tasks, including browsing the web and connecting to streaming music accounts so that it can act as a music player. There are some points to bear in mind when using Facebook Portal:

Beware

You cannot take a screenshot of your Portal's screen. This is a security feature so that people cannot capture the screen when you are having a video chat with them.

- Facebook Portal links to your Facebook account, and this is required in order to use your Portal.

- Video calls can be made to your Facebook contacts, and also WhatsApp contacts.

- Recipients of a Portal video call do not need to have a Portal device themselves; Portal calls can be made to anyone with the Messenger or WhatsApp apps.

- The Portal camera can track your position, so you can move around during a video call and the camera will follow you.

Beware

If you remove the power adapter at any point, your Portal will turn off.

- Your Portal has to be connected to a power source in order for it to operate.

- Portal's power adapter has a tubular connector that attaches to Portal and that also doubles up as a stand for it.

Don't forget

Portal has a thick border around the body of the device, which also helps support it either horizontally or vertically.

Around your Portal

The body of your Portal contains the elements that provide its functionality. These include:

- The Portal camera, which is located in the top left-hand corner as you are looking at the screen. When the camera and  microphone are activated (i.e. for a video call) a green indicator is displayed to the right of the camera.

- Use this button on the top of your Portal, to turn video and audio functionality On or Off.

- Move the button one space to the left to turn Off the video functionality.

- Move the button another space to the left to turn Off the video and audio functionality.

- When the power adapter is connected to your Portal, and plugged in, the startup process begins.

Beware

The indicator light turns red if the camera and microphone have been turned off with the buttons on the body of your Portal.

Beware

The camera and microphone of your Portal are always on, ready for use. However, this means that they are active even when your Portal is not being used. If you feel uncomfortable about this, turn off the power supply when you are not using your Portal.

Getting Started with Portal

When you first plug in your Portal, there are a few initial steps to get it up and running:

1 Select a system language to use for your Portal and tap on the **Next** button

2 Tap on an available **Wi-Fi** network

Tap on the eye icon in Step 3 to show or hide the password as you are entering it, to ensure that you have entered it correctly.

3 Enter the password for the Wi-Fi router and tap on the **Join** button

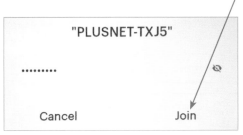

4 The selected Wi-Fi network is connected, indicated by a tick symbol next to it

5 Tap on the **Next** button

6 Tap on the **Continue** button to go to the next stage of setting up your Portal

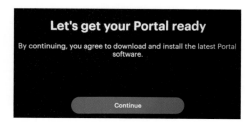

7 If there is a software update for your Portal, this will be installed automatically

Portal's software will be updated if there has been a newer version released since your Portal was manufactured.

Setting Up Portal

Once any software updates have been downloaded for your Portal, as shown in Step 7 on page 171, your Portal is ready for the setup process:

1 Tap on the **Next** button to start using, or creating, your Facebook account with your Portal

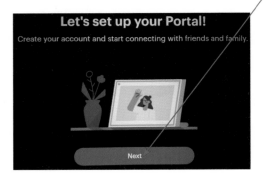

Hot tip

The default names in Step 3 are a good option if you have more than one Portal, in different rooms in the house.

2 Tap on the **Continue** button to accept the **Terms of Service**

3 Tap on a location to apply that as your Portal's name, or tap on the **Custom Name...** option

4 Enter a custom name for your Portal and tap on the **Save** button

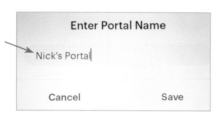

5 The name is applied to your Portal. Tap on the **Next** button

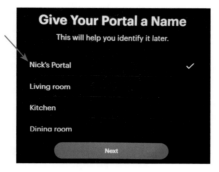

6 Select an account to link to your Portal (Facebook or WhatsApp) and tap on the **Continue** button

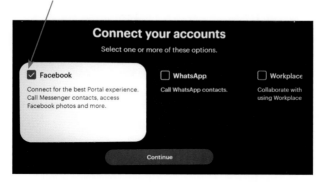

If you select to add your Facebook account in Step 6, you can always add a WhatsApp account too at a later time. See pages 181-182 for details.

173

7 Use the code that is displayed and enter it into the website at **www.facebook. com/device**, or tap on the **Use Facebook Password Instead** option, to access your Facebook account

...cont'd

8 For a Facebook account, enter a username and tap on the **Return** button on the keyboard

9 Enter a password for the account and tap on the **Return** button on the keyboard

10 Confirmation for connecting to your Facebook account is displayed. Tap on the **Next** button

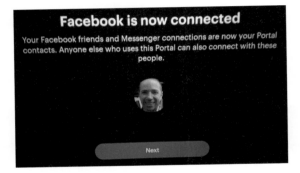

11 Tap on contacts from Facebook to add them to favorites within your Portal, so that you can connect with them quickly. Tap on the **Continue** button

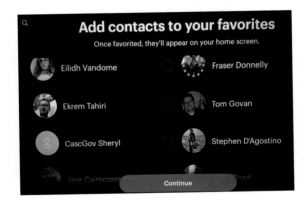

If you want to prevent some of your Facebook contacts from being able to contact you via your Portal, tap on the **Hide Certain Contacts** button after Step 11 and select the people, as required.

12 Apply options for adding streaming-music options to your Portal, and photos from your Facebook account

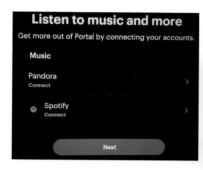

13 At the end of the setup process, check this box On or Off to receive Portal updates and notifications. Tap on the **Continue** button

14 Tap on the **Explore Home** button to start using your Portal

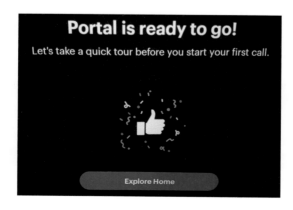

Two other options during the setup process are for adding functionality for an Alexa smart speaker and for using the phrase "Hey Portal" to activate your Portal. Both of these options can be added, and edited, in the **Assistants** section of the Portal settings.

Around the Home Screen

After your Portal has been set up, the Home screen is displayed. This is the screen that you will return to when you are not making a video call with your Portal, or using any other of its functions. To use the Home screen:

1 Tap on the **Call Demo** button to access options for a test video call

2 The demo call is connected. Both of the video feeds are your own at this point

3 Tap on the screen to access the video control buttons. Tap on this button to end the demo call and return to the Home screen

4 Swipe from right to left on the Home screen in Step 1 to view the other available Home screens

5 Tap on the buttons on the additional Home screens to access more functions, such as using a web browser on your Portal to access the web

6 Tap on the **All Contacts** button on the Home screen to access people for video chats. See pages 180-182 for details

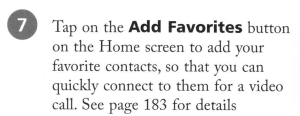

7 Tap on the **Add Favorites** button on the Home screen to add your favorite contacts, so that you can quickly connect to them for a video call. See page 183 for details

8 Tap on this button at the top of any Home screen to activate the screen saver option of different photos

Don't forget

To see more details about the Portal settings, which can be accessed in Step 4, see pages 178-179.

Don't forget

Swipe from right to left on the second screen in Step 4 to access a third one, which contains the **Help** option.

Don't forget

Tap on the **Photos** button in Step 4 to add photos from your Facebook account or view ones that you have already added.

Portal Settings

The Portal settings can be used to customize it in a number of ways. To access and use the settings:

1 On the Home screen, swipe from right to left and tap on the **Settings** button

2 The main Settings categories are displayed in the left-hand panel

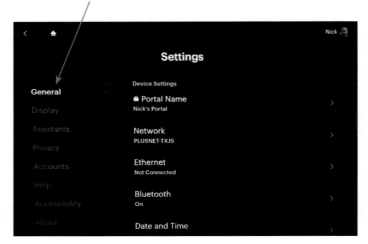

3 Tap on a category in the left-hand panel to view its options in the main window

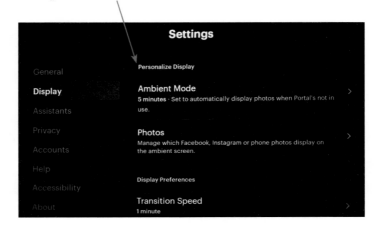

General settings
These can be used for options including: changing your Portal's name; connecting to a Wi-Fi network; connecting a Bluetooth device; setting the date and time; setting the device language; and sound settings.

Display settings
These can be used for options including: displaying photos when your Portal is not in use; managing photos used on your Portal; turning the screen off automatically after a period of inactivity; and adjusting the screen brightness and color.

Assistants settings
These can be used for setting up the "Hey Portal" function for activating your Portal with voice commands, and setting up an Amazon Alexa smart speaker to interact with your Portal.

Privacy settings
These can be used for options including: displaying your Facebook status to your contacts; hiding contacts on Messenger and WhatsApp; and locking the screen, so that it has to be unlocked with a passcode.

Accounts settings
These can be used to add more users to your Portal (using their Facebook account details) and linking to other online accounts, such as Spotify for playing music on your Portal.

Help settings
These can be used for giving feedback about using your Portal and also accessing the Portal Help app, which contains a wealth of information about using your Portal.

Accessibility settings
These can be used for options including: changing the font size used on your Portal; changing the screen colors to make text easier to read; changing the touch-and-hold delay when you tap on items on the screen; using the TalkBack screen reader for reading what is on the screen; and using closed captions (subtitles) during a video call.

In the Display settings, the **Screen Off** option can be used to set a time after which the Portal screen will turn off if no motion is detected by the camera. However, this does not turn off your Portal and it will be reactivated if the camera detects any motion, even if you are unaware of it being activated again.

Your own WhatsApp account can be linked to your Portal from the Accounts settings, in addition to the option for linking it, as shown on pages 181-182.

Portal Contacts

Facebook Portal can be used to make video calls to your Facebook contacts, and also contacts that you have in the WhatsApp app, which can be linked to your Portal, if you have not already done so. To access all of your available contacts on your Portal:

Don't forget

The **Suggested** contacts in Step 2 are the same as those you will see on your Facebook page.

1 On the Home screen, tap on the **All Contacts** button

2 By default, **Suggested** contacts are displayed. Tap on the video icon to start a video chat with someone

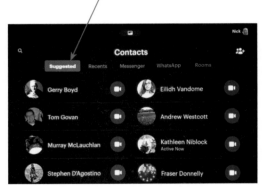

3 Tap on the **Messenger** button on the top toolbar to view your Facebook contacts. Tap on the video icon to start a video chat with someone

Connecting to WhatsApp

WhatsApp is another communication app that is used primarily for text chats, but it is also highly effective for video chats. WhatsApp is owned by Facebook and it can be linked to your Portal so that your WhatsApp contacts can be used for video chats with your Portal too. To do this:

1 Access the **All Contacts** section, as shown on the previous page, and tap on the **WhatsApp** button on the top toolbar

2 If you have not already added your WhatsApp contacts, tap on the **Connect WhatsApp** button

3 Use the code that is displayed and enter it into the website at **www.facebook. com/device**, or tap on the **Use Facebook Password Instead** option

4 Check this box **On** and tap on the **Continue** button

The WhatsApp app is usually used on a smartphone.

...cont'd

Don't forget

A QR code stands for Quick Response code. It is a type of barcode that can contain information required for specific tasks that can be carried out by a user once the code has been scanned.

5 A QR code is displayed. This is used to verify your WhatsApp account with your Portal.

Open WhatsApp on your smartphone, access the **WhatsApp Web** option, tap on the **Scan this QR code** option and point your device's camera at the code to scan it

6 When the confirmation window appears, tap on the **Done** button

7 Your WhatsApp contacts are displayed, under the **WhatsApp** button

Adding Favorite Contacts

The family and friends who you video call most frequently can be added as Favorites on your Portal so that they are only ever a tap or two away. To do this:

1 On the Home screen, tap on the **Add Favorites** button

2 Select either your Messenger or WhatsApp contacts page

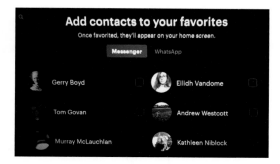

3 Check this button **On** to select someone, and tap on the **Add** button

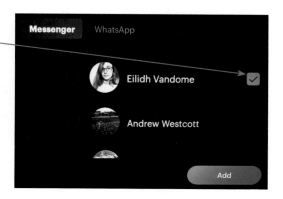

Tap on a favorite contact to access options for starting a video call or an audio call, or sending them a text message via Messenger.

4 The person is added below the **Favorites** heading on the Home screen. Tap on the **Add Favorites** button to add more people as favorites, as required

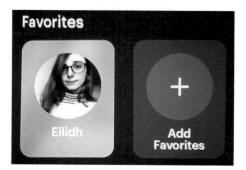

Making a Portal Call

To make a video call to your contacts on your Portal (either Messenger or WhatsApp contacts):

1 Access the required contact, from either the **All Contacts** button or the **Favorites** button on the Home screen

2 Tap on the video icon next to the selected person

3 When the call is accepted, the recipient's video feed is in the main window and your video feed is a thumbnail in the top left-hand corner

Hot tip

Tap on this button on your own video feed in Step 4 to minimize the thumbnail at the side of the screen:

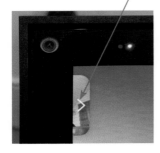

4 Tap on your own video feed and tap on this button to switch windows

5 Your video feed becomes the main one. Reverse the process to minimize your own video feed

6 Tap on the main window to access the video control buttons. Use the buttons to, from left to right: add more people to the call; access additional media options to add to the call; turn your video feed on or off; mute or unmute your microphone; and end the current call

7 The media options can also be accessed by swiping up on the screen when the video controls are visible

8 Tap on the **Story Time** button to view and read a story on your Portal. As you read the text, different animations appear on the screen, with your video feed

The **Story Time** option is an excellent one for reading stories to grandchildren, if you cannot be with them in person.

...cont'd

Adding a room

Rooms can be created on your Portal, to enable groups of people to have video chats on a regular basis. Chats can be done with everyone in a room, or just whoever is available. To create a room on your Portal:

Rooms on your Portal can only be created with contacts on Facebook, but not WhatsApp ones.

1. Access the **All Contacts** section and tap on the **Rooms** button on the top toolbar

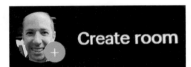

2. Tap on the **Create room** button

3. Tap on the **Room activity** option to enter a description for the room

4. Tap on the **Who can discover and join?** option to specify who can join the room

5. Tap on the **Settings** button to apply a range of settings for the room and then tap on the **Next** button

6. Select how your room can be viewed and tap on the **Create Room** button

L

M

N

O